Lennart Merkert

Optimal Scheduling of Combined Heat and Power Generation Considering Heating Grid Dynamics

Karlsruher Beiträge zur
Regelungs- und Steuerungstechnik
Karlsruher Institut für Technologie

Band 08

Optimal Scheduling of Combined Heat and Power Generation Considering Heating Grid Dynamics

by
Lennart Merkert

KIT Scientific Publishing

Karlsruher Institut für Technologie
Institut für Regelungs- und Steuerungssysteme

Optimal Scheduling of Combined Heat and Power Generation Considering Heating Grid Dynamics

Zur Erlangung des akademischen Grades eines Doktor-Ingenieurs von der KIT-Fakultät für Elektrotechnik und Informationstechnik des Karlsruher Instituts für Technologie (KIT) genehmigte Dissertation

von Dipl.-Ing. Lennart Merkert

Tag der mündlichen Prüfung: 10. Juli 2020
Hauptreferent: Prof. Dr.-Ing. Sören Hohmann
Korreferent: Prof. Dr.-Ing. Hendrik Lens

Impressum

KIT Scientific Publishing

Karlsruher Institut für Technologie (KIT)
KIT Scientific Publishing
Straße am Forum 2
D-76131 Karlsruhe

KIT Scientific Publishing is a registered trademark of Karlsruhe Institute of Technology.
Reprint using the book cover is not allowed.

www.ksp.kit.edu

Print on Demand 2021 – Gedruckt auf FSC-zertifiziertem Papier

ISSN 2511-6312
ISBN 978-3-7315-1056-7
DOI 10.5445/KSP/1000125176

Preface

This PhD thesis was developed while working at ABB Corporate Research Germany in Ladenburg, Germany in collaboration with the Institute for Control Systems (IRS) at Karlsruhe Institute of Technology (KIT).

First of all, I would like to thank Prof. Dr.-Ing. Sören Hohmann for his guidance, for challenging my research hypotheses and for giving me the opportunity to become an external PhD student at IRS. In addition, I am very grateful to Prof. Dr.-Ing. Hendrik Lens for taking the time to act as second examiner.

I would like to thank all my colleagues at ABB Corporate Research for great collaborations, interesting discussions and a nice working environment. Special thanks go to my master students and interns Julia Brockschmidt, Julian Bauer, Ashvar Abdoul Haime and Tobias Lorenz who contributed with many questions, fruitful discussions and interesting results. Furthermore, many thanks to Jan Schlake and Susanne Schmitt for proofreading this manuscript. This work would not have been possible without the support of my managers Veronica Biagini and Kim Listmann keeping me out of many every day tasks and supporting me as much as possible in their limited time.

Many thanks to all PhDs and Post-Docs at Institute for Control Systems at Karlsruhe Institute of Technology for their warm welcome and many interesting discussions, especially with Jona Maurer who worked on related topics.

In addition, I would like to thank Andreas Kohler and Annette Vogt from Stadtwerke Kiel for giving me many insights in real-world heating grid operations and providing detailed measurement data.

Without Pedro Castro from University of Lisboa, Portugal, the optimization approach using the hybrid discrete-continuous time grid would not have been possible. Thanks a lot to Pedro for this very fruitful and inspiring collaboration.

Last but not least I would like to thank my parents and my brother for raising my curiosity and supporting my academic education as well as Britta for standing by my side through heights and depths.

Heidelberg, February 2020

To Britta

Abstract

Combined heat and power plants (CHPs) enable an efficient co-generation of heat and electricity and thus are crucial for a resource efficient energy supply. As the share of renewable generation is increasing in electric grids, the operation of CHPs gets more and more challenging. The traditional heat-driven operation of CHPs is not suitable to react to the volatile availability of renewable energy, to variable electric demand as well as to fluctuating electricity prices. However, flexible operation of CHPs requires thermal storage. As an alternative to dedicated thermal storage tanks, the water inside a heating grid can be used as thermal storage if grid dynamics are considered in operations planning. It is very complex to find good or, ideally, optimal operation schedules for CHPs when the dynamics of a heating grid are included. These dynamics are dominated by a variable-dependent time delay, which results in a non-convex problem formulation. Hence, common optimization approaches reach their limits and do not find the global optimum when the thermal dynamics of a heating grid are considered. This thesis presents the following new methods to nevertheless find optimal schedules for operation of CHPs considering heating grid dynamics: i) the first method solving problems with variable-dependent time delays to global optimality by proposing a novel outer approximation of the pipe outflow temperature in a primal-dual global optimization scheme, ii) a method introducing a hybrid discrete-continuous time grid and discretized temperatures to enable an accurate representation of the variable-dependent time delays in a mixed-integer linear program and iii) an approach allowing a measurement based identification of the heating grid dynamics that enables an easy application to real world grids.

Kurzfassung

Kraft-Wärme-Kopplung (KWK) ermöglicht eine effiziente Erzeugung von Wärme und Elektrizität und spielt somit eine zentrale Rolle für eine ressourcenschonende Energieversorgung. Mit stetig wachsender Einspeisung erneuerbarer Energieträger ins Stromnetz steigt die Volatilität am Strommarkt und regelbare Erzeuger müssen flexibel auf diese Schwankungen reagieren können. Die traditionell wärmegeführte Betriebsweise von KWK-Anlagen ist deshalb nicht mehr zeitgemäß. Für einen flexibleren Betrieb von KWK-Anlagen sind jedoch Wärmespeicher notwendig. Nicht nur in Speichertanks, sondern auch im Wärmenetz selbst kann Wärmeenergie gespeichert werden. Hierzu muss die Dynamik des Wärmenetzes in der Einsatzplanung der KWK-Anlagen berücksichtigt werden. Die Modellierung und Optimierung dieser thermischen Dynamik ist jedoch komplex, da sie auf einer variablen-abhängigen Verzögerung beruht. Diese variablen-abhängige Verzögerung führt zu einer nicht-konvexen Problemstellung, für die herkömmliche Optimierungsverfahren nicht ausgelegt sind. Um trotzdem optimale Lösungen zu erreichen, werden in dieser Arbeit folgende neue Methoden zur optimalen Einsatzplanung von KWK-Anlagen unter Berücksichtigung der Dynamik von Wärmenetzen entwickelt: i) die erste Methode, die Probleme mit variablen-abhängigen Verzögerungen global optimal löst, indem eine neue äußere Abschätzung der Rohraustrittstemperatur als duales Problem in einem iterativen, globalen Optimierungalgorithmus eingesetzt wird, ii) eine Methode, die mit Hilfe einer Diskretisierung der möglichen Vorlauftemperaturen sowie einem hybriden diskret-kontinuierlichen Zeitstrahl eine exakte Formulierung der Verzögerung als gemischt-ganzzahliges Problem erlaubt und iii) eine Methode, die eine messwertgestützte Modellidentifikation der Wärmenetzdynamik ermöglicht und somit einfach in realen Netzen eingesetzt werden kann.

Contents

List of Figures

List of Tables

Abbreviations and Nomenclature

Abbreviations

Abbreviation	Description
CHP	combined heat and power plant
MILP	mixed-integer linear program
MINLP	mixed-integer non-linear program
NLP	non-linear program

Nomenclature

Sets

Symbol	Description
$\mathbb{N}$	natural numbers
$\mathbb{R}_{\geq 0}$	set of real, non-negative numbers
S_c	set of all possible combinations for running a CHP in current and past time slots
S_d	set of demands (loads) considered
S_g	set of generators considered
S_i	set of n volume elements
S_j	set of n_p sub-volumes
S_k	set of possible values of a digit (0 to 9)
S_l	set of temperature levels
S_t	set of time points in optimization horizon
$S_{t>1}$	set of time points in optimization horizon without first time point
$S_{t<n_t}$	set of time points in optimization horizon without last time point
S_t^{fx}	set of fixed time points in optimization horizon
S_τ	set of all possible (discretized) time delays
$\mathbb{Z}$	integer numbers

Parameters

Symbol	Description
C_i^{const}	constant cost for running CHP i
$C_i^{el,var}$	variable cost for producing electricity with CHP i
$C_i^{heat,var}$	variable cost for producing heat with CHP i
$Comb_i$	a specific combination i for running a CHP in current and past time slots
c_p	specific heat capacity of transport medium, assumed to be constant
d	diameter of pipe
d_t^1, d_t^2	length of volumes at temperatures $T_{t-1}^{in}, T_{t-2}^{in}$ leaving the pipe in time slot t
k_V	heat loss factor of pipe
L	length of pipe
$M[l,t]$	matrix used to model thermal dynamics in delay matrix approach
$M^\star$	large number used for linking binary with real variables in MILP
$\dot{m}^L$	minimum mass flow of transport medium in pipe
$\dot{m}^U$	maximum mass flow of transport medium in pipe
n_p	number of sub-volumes in multiparametric delay modeling
n_p^*	a specific number of sub-volumes
n_t	number of time points
$n_{T,t}$	number of possible temperature levels in pipe in time slot t
n	number of volume elements for finite volumes and hybrid pipe model
P_i^L	minimum electric power output of CHP i
P_i^U	maximum electric power output of CHP i
$price_t^{el,buy}$	market price for purchasing electricity in time slot t
$price_t^{el,sell}$	market price for selling electricity in time slot t
Q_i^L	minimum heat power output of CHP i
Q_i^U	maximum heat power output of CHP i
p_t^{demand}	electric demand in time slot t
q_t^{demand}	heat demand in time slot t
$q^{demand,max}$	maximum heat demand in optimization horizon
s^1, s^2	share of volume at $T_{t-1}^{in}, T_{t-2}^{in}$ leaving the pipe in a time slot
$s^{m,1}, s^{m,2}$	combine s^1, s^2 with mass flow
T^{amb}	ambient temperature
T^{inside}	temperature of transport medium inside a pipe
T^L	minimum possible (supply) temperature
T^{return}	temperature of transport medium in return network, assumed to be constant
T^U	maximum possible (supply) temperature
Δt	length of time slot
v^U	maximum velocity of transport medium in pipe

Symbol	Description
$\alpha_{\tau,t}$	weight factors used in node method
α_{loss}	linear loss factor used to model thermal losses with delay matrix approach
$\alpha_1, \alpha_2, \alpha_3$	parameters for limitation of CHP feasible operation region
$\beta_1, \beta_2, \beta_3$	parameters for limitation of CHP feasible operation region
ρ	density of transport medium, assumed to be constant
τ^{min}	minimum transport time for highest possible velocity in a pipe
τ^{max}	maximum transport time from entering to leaving pipe

Variables

Symbol	Description
$b_{\tau,j,t}$	binary variable describing discretized time delay τ of sub volume j in time slot t in multiparametric delay modeling
δ^+, δ^-	auxiliary variables to represent min and max function
$\dot{m}_t$	mass flow in time slot t (a parameter in delay matrix approach)
$n_{i,t}$	binary variable for commitment status of CHP i in time slot t (0: off / 1: on)
$p_{i,t}$	electric power output of CHP i in time slot t
p_t^{buy}	electric power purchased at market in time slot t
p_t^{sell}	electric power sold to market in time slot t
$\dot{Q}_{i,t}$	change of inner energy of volume element i in time slot t
$\dot{Q}_{i,t}^{loss}$	change of inner energy of volume element i in time slot t based on thermal losses
$\dot{Q}_{i,t}^{convection}$	change of inner energy of volume element i in time slot t based on convective transport from and to neighboring volume elements
$q_{i,t}$	heat power output of CHP i in time slot t
$\hat{q}_{i,t}$	simulation result for heat power output of CHP i in time slot t
q_t^{charge}	thermal energy charged to or discharged from heating grid in time slot t
q_t^{in}	thermal energy flowing into a pipe in time slot t
q_t^{loss}	thermal losses in time slot t
q_t^{out}	thermal energy flowing out of a pipe in time slot t
SOC_t^{heat}	state of charge of thermal energy in time slot t
$T_{i,t}^{av}$	average temperature of volume element i in time slot t
T_t^{in}	inflow temperature in time slot t
$\widetilde{T}_{t-\tau}^{in}$	delayed inflow temperature reduced by thermal losses occurring in transport time τ
T_t^{out}	outflow temperature in time slot t

Symbol	Description
$T_{t,j}^{out}$	outflow temperature of sub volume j in time slot t
T_t^{supply}	supply temperature in time slot t
v_t	velocity of transport medium in pipe in time slot t
Δx_t	propagation of transport medium in pipe in time slot t
$\Delta x_{j,t}$	span of sub-volume j in time slot t
$y_{t,i}$	binary variable indicating if current and and past mass flows in time slot t match mass flow combination i
τ_t	transport time from entering to leaving a pipe at time t
τ_t^{flex}	remaining variable part of transport time with hybrid pipe model in time slot t

Additional variables and parameters of the hybrid time grid approach

Symbol	Description
$F_{l,t}^C$	volume at temperature level l produced in time slot t and entering pipe
$F_{l,t}^D$	volume at temperature level l consumed in time slot t and leaving pipe
f^L	lower bound of volume entering or leaving pipe
f^U	upper bound of volume entering or leaving pipe
L_t	duration of time slot t (continuous)
$LC_{l,t}$	left (volume) coordinate in pipe of the volume at temperature level l in time slot t
$LC_{l,t}^{end}$	left (volume) coordinate in pipe of the volume at temperature level l at the end of time slot t
lc_l^0	left (volume) coordinate in pipe of the volume at temperature level l at beginning of the optimization horizon
l	(discrete) temperature level
$RC_{l,t}$	right (volume) coordinate in pipe of the volume at temperature level l in time slot t
$RC_{l,t}^{end}$	right (volume) coordinate in pipe of the volume at temperature level l at the end of time slot t
rc_l^0	right (volume) coordinate in pipe of the volume at temperature level l at beginning of the optimization horizon
t_t	time of time point t (continuous)
tfx_t	fixed time points
T_l^{supply}	temperature of temperature level l
$V_{l,t}$	volume at temperature level l in pipe in time slot t
v	overall volume of pipe
v_l^0	volume at temperature level l in pipe at beginning of the optimization horizon

Symbol	Description
$X^C_{l,t}$	binary indicating if temperature level l is produced in time slot t
$X^D_{l,t}$	binary indicating if temperature level l is consumed by load in time slot t
$X^P_{l,t}$	binary indicating if temperature level l is inside the pipe in time slot t
$ZC_{l,t}$	real variable allowing to reset the coordinates $LC_{l,t}$ and $RC_{l,t}$ in case temperature level l is not inside the pipe

Variables and parameters of multiparametric disaggregation

Symbol	Description
l	position of digit
k	value of digit
P	upper bound of position of digit
p	lower bound of position of digit
$w_{i,j}$	product (real variable)
Δw_{ij}	slack variable for product term
x_i, x_j	multipliers (real variables)
x_i^L	lower bound of multiplier x_i
x_i^U	upper bound of multiplier x_i
$\hat{x}_{ijkl}$	real variable for multiplication per digit l
Δx_j	slack variable for multiplier x_j
z_{jkl}	binary variables replacing x_j

1 Introduction

The reduction of green house gas emissions and an efficient use of energy resources are a major concern in today's societies. District heating grids play an important role for a sustainable energy supply as they enable efficient co-generation of heat and electricity in combined heat and power plants (CHPs) as well as the use of thermal energy normally dissipated to the environment e.g. from industrial production processes or waste incineration. District heating grids transport thermal energy from heat sources to heat sinks using mostly water as transport medium. This water is heated at generation sites to a supply temperature chosen by the heating grid operators. Then the hot water is transported via supply pipes to heat consumers. They cool down the hot water to a return temperature and the water is transported back to the generation sites via separate return pipes.

Worldwide there are about 80,000 heating grid systems supplying about 14 EJ thermal energy per year of which 6,000 are located in Europe supplying about 2.5 EJ thermal energy each year [Wer17]. The operators of these heating grids are facing multiple challenges today. On the one hand, they need to supply the thermal demand as resource-, energy- and cost-efficiently as possible. On the other hand, energy and especially electricity markets are becoming more and more volatile due to a rising share of renewable power generation. Hence, the traditional heat-driven operation strategy for CHPs reaches its limits and operators need to consider intra-day fluctuations of prices and availability of electricity. Thus, a closely coupled coordination of co-generation of different energy forms is needed, which is one of the central topics of 4th generation district heating [LWW+14].

In particular, a better integration of different energy sectors provides benefits for renewable power integration into electric grids. In electric power systems, generation needs to equal demand at all times. As renewable generation is not fully controllable, energy storage is of major importance to balance demand and supply in power grids with a high share of renewable generation. Storage units that are able to store energy for minutes, hours, days and months are required to stabilize these future grids. Pumped-hydro storage, the most affordable form of electric energy storage, as well as thermal energy storage tanks are both very suitable for storage cycles of up to a few days. However, investment costs for thermal energy storage tanks are orders of magnitude cheaper than for pumped-hydro storage [LØC+16]. Hence, using thermal energy storage instead of electric storage reduces investment costs importantly. In addition, thermal energy can directly be stored in a heating grid for a few hours thanks to its dynamic behavior [Gro12]. Thus, every heating grid offers thermal storage capacity without any investment costs, if the thermal dynamics of the grid are considered in operations planning.

However, using this inherent grid storage in operations planning is a very complex task, as the thermal dynamics of a heating grid are driven by a mass-flow dependent transport time, which leads to a time delay in the arrival of an increase of temperature from the beginning to the end of a pipe. When applying optimization approaches to operations planning problems for district heating grids, this variable-dependent delay results in a non-convex problem formulation. Due to its importance for future energy systems, several publications have addressed this operational optimization of heating grids. As shown in Chapter 2, iterative and sequential solution approaches have been proposed to cope with the non-convex variable-dependent time delay. However, none of these approaches prove global optimality of their solution. Thus, it is unknown if they leverage the full potential of the heating grid as thermal energy storage capacity. In addition, the algorithms that have been proposed in the literature are only rarely used in real world applications as they are complicated to implement and use.

Therefore, this work addresses the following research questions:

- Can we develop an algorithm which reliably finds the global optimal solution for operational optimization of combined heat and power plants whilst considering the dynamics of heating grids?
- Can we find exact and easy to engineer algorithms which offer fast solution times and thus are applicable for real world heating grids?

To tackle the first question, Chapter 3 presents "multiparametric delay modeling" the first approach reaching global optimality for operational planning of heating grids. "Multiparametric delay modeling" introduces binary variables that represent the current transport delay and are used to derive an outer-approximation of the outflow temperature of a pipe. Chapter 4 and Chapter 5 address the second question, facilitating optimization of heating grids in real-world operations planning. The hybrid-time grid approach presented in Chapter 4 focuses on a fast to solve but still very accurate model. It adapts optimization models recently developed for pipeline scheduling in the petrochemical industry by introducing discrete temperature levels and a hybrid discrete-continuous time grid. The main focus of the delay matrix approach presented in Chapter 5 lies on an easy and fast parameterization of the optimization model. The model dynamics are represented in a static delay matrix which can be calculated using historic temperature and flow measurements. Chapter 6 then compares the optimization approaches presented in this thesis to each other as well as with a literature approach using small example heating grids. In addition, a real-world case study is presented that demonstrates the scalability and benefits of the approach of Chapter 5. Finally, Chapter 7 summarizes and discusses the results of this work.

2 State of the art

As mentioned in the introduction, several past publications address the optimization of heating grids. This chapter gives first an overview on the state of the art for operational optimization of heating grids. Second, theory on global optimization of non-convex problems is introduced with focus on non-convex problem classes that are required to describe thermal dynamics of heating grids. In this literature review it becomes apparent that current optimization theory does not provide suitable methods for global optimization of heating grid dynamics. Third, Section 2.3 introduces pipeline scheduling for crude oil products which models similar flow dynamics as found in heating grids.

2.1 Operational optimization of district heating grids

Trying to find optimal set points for operation of district heating grids has been studied in many past publications. Common goals in operational optimization of district heating grids are the reduction of green house gas emissions as well as the reduction of operational cost. However, different optimization problems aiming at these goals can be distinguished in literature: i) If multiple heat production units are part of the grid, a key question is to find their schedules determining which unit produces when [SP08]. ii) Finding the optimal supply temperature is of high interest, which ensures security of supply for all consumers while reducing thermal and hydraulic losses [AMP14, VTD17]. A combination of both problems is possible, resulting in a fully dynamic optimization of the heating grid [BBR95].

In this thesis, the main focus lies on scheduling of production units. However, the global optimization approach presented in Chapter 3 offers a fully dynamic optimization.

2.1.1 Scheduling of CHPs without consideration of thermal dynamics

For scheduling of CHPs, district heating grid dynamics are very often neglected and a static heat load forecast is assumed. With these assumptions, it is common to use a mixed-integer linear programming (MILP)[1] discrete time unit commitment model like, e.g., in [Dot97, NNM00, ROGJ17]. This section introduces such a basic discrete time MILP model for scheduling of CHPs without consideration of thermal dynamics. Existing applications and extensions of this model are reviewed at the end of this section.

With the discrete time representation we get a set S_t of time slots t of equal duration. As input parameters do not vary within each time slot, model accuracy is not reduced. In addition, variables are constant within every time slot t. All CHPs are collected in set S_g. For every CHP i in each time slot t, we introduce non-negative real variables representing the electric power output $p_{i,t}$ and the thermal power output $q_{i,t}$ as well as the running status of the unit $n_{i,t}$ being a binary variable (0: off / 1: on). Lower limits of power and heat output are P_i^L and Q_i^L and the respective upper limits of power and heat output are P_i^U and Q_i^U. To enforce that the CHP only produces power and heat when it is running ($n_{i,t} = 1$) we get

$$n_{i,t} \cdot P_i^L \leq p_{i,t} \leq n_{i,t} \cdot P_i^U \qquad \forall i \in S_g, \forall t \in S_t, \tag{2.1}$$

$$n_{i,t} \cdot Q_i^L \leq q_{i,t} \leq n_{i,t} \cdot Q_i^U \qquad \forall i \in S_g, \forall t \in S_t. \tag{2.2}$$

Electric power output $p_{i,t}$ and heat power output $q_{i,t}$ of a CHP are not independent. This relationship is influenced by the used generation technology. Gas turbines as well as gas engines usually have a linear connection between heat and electric output and hence a fixed power-to-heat ratio, whereas for CHPs with extraction condensing turbine the power-to-heat ratio is flexible and limited by a capability region as shown in Figure 2.1 [Cer02]. The feasible region of a CHP i with extraction condensing turbine shown in Figure 2.1 can be described with inequalities

$$\begin{aligned} p_{i,t} &\leq \alpha_1 - \beta_1 \cdot q_{i,t} && \forall t \in S_t, \\ p_{i,t} &\geq \alpha_2 - \beta_2 \cdot q_{i,t} && \forall t \in S_t, \\ p_{i,t} &\geq -\alpha_3 + \beta_3 \cdot q_{i,t} && \forall t \in S_t \end{aligned} \tag{2.3}$$

with α_1, α_2, α_3, β_1, β_2 and β_3 being parameters describing slope and offset of the limiting lines.

Assuming that electric demand p_t^{demand} and thermal demand q_t^{demand} are known, energy balances for heat and electricity for every time slot t can be formulated. Because CHPs are usually connected to a large electricity grid, it is assumed that electric energy can be sold and purchased. The volume of electricity sold in time slot t is p_t^{sell}, whereas the volume of electricity bought in time slot t is p_t^{buy}, both being non-negative real variables. If no energy storage like battery or pumped-hydro storage is installed,

[1] MILPs are optimization problems with real and integer (discrete) variables. The problem formulation consists of a linear objective function to be minimized or maximized as well as linear equality and inequality constraints linking real and integer variables. In many cases the integer variables are binary taking only two values. There are many off-the-shelf solvers to solve MILPs such as [BEN05, Gur16, FRV+18].

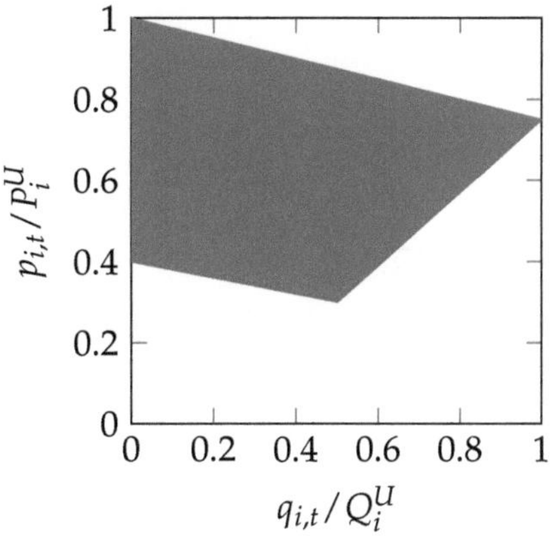

Figure 2.1: Feasible operation region of a CHP i with extraction condensing turbine linking normalized electric output $p_{i,t}/P_i^U$ and normalized thermal output $q_{i,t}/Q_i^U$ following [Cer02].

electric energy needs to be produced at the same time it is consumed. Thus, the electric power balance becomes

$$p_t^{demand} = \sum_{i \in S_g} p_{i,t} + p_t^{buy} - p_t^{sell} \qquad \forall t \in S_t. \tag{2.4}$$

For thermal energy there are usually no open markets enabling hourly sales and purchases, leading to

$$q_t^{demand} + q_t^{loss} = \sum_{i \in S_g} q_{i,t} \qquad \forall t \in S_t. \tag{2.5}$$

In this thermal energy balance, heat losses q_t^{loss} of the heating grid are accounted for as well.

In unbundled electricity markets (like all electricity markets in Europe) utilities producing electricity do not need to consider transmission losses in the electric grid, as this is the responsibility of grid operators. They purchase energy at the energy markets to compensate losses in transmission and distribution grids. Thus, for operations planning of a power producer grid losses are not relevant and do not appear in the electric power balance in (2.4).

To complete the MILP formulation for scheduling of CHPs, the objective function is the last missing piece. As mentioned in Section 2.1, major goals for operation of district heating grids are the reduction of CO_2 emissions and the reduction of cost. If CO_2 emissions are considered in the generation cost, both goals can be combined by minimizing the objective function

$$\min \sum_{t \in S_t} \left(price_t^{el,buy} \cdot p_t^{buy} - price_t^{el,sell} \cdot p_t^{sell} \right) + \sum_{t \in S_t} \sum_{i \in S_g} \left(C_i^{const} \cdot n_{i,t} + C_i^{el,var} \cdot p_{i,t} + C_i^{heat,var} \cdot q_{i,t} \right). \quad (2.6)$$

Here $price_t^{el,buy}$ and $price_t^{el,sell}$ are time varying prices for buying and selling electricity and C_i^{const} is the cost for running CHP i in one time slot. $C_i^{el,var}$ and $C_i^{heat,var}$ are the variable cost for the production of electricity (el) and heat of CHP i.

Existing applications and extensions of the basic CHP scheduling model

Many publications use a scheduling model as introduced above. For example, in [Dot97] the author proposes an economic dispatch model for CHPs which allows flexible generation by extending the model with thermal storage tanks. Time varying cost of generation and time varying prices for sales to the energy markets are considered in scheduling of decentralized CHPs in [NNM00]. As well considering decentralized CHPs with thermal storage tanks, experiences from a real-world online optimization are shared in [FKBV13]. Here, an intra-day optimization reacts to unforeseen variations in generation or load. Another work investigates how future electricity prices can influence the operations strategy of CHPs [ROGJ17]. A review of different methods for short term operations planning of CHPs without consideration of grid dynamics is given in [SP08]. Being not intended for large heating grids, the algorithm proposed in [LM17] as well does not consider time delays in temperature propagation. It uses a Taylor series expansion around a given working point to linearize the remaining non-linear heating grid dynamics.

2.1.2 Thermal dynamics of heating grids: the node method

The model and the approaches presented in the previous section do not consider grid dynamics. To consider dynamics of a heating grid in simulation or optimization, the node method [BBR95] is a very common approach of modeling. Besides many academic works (e.g. [SFT+05, LWS+16, ZFZC18]), commercial simulation tools rely on it [Sol06] and simulations show that it achieves a better performance than other modelling approaches [Trö99].

Thus, this section introduces the basic principles of thermal dynamics of heating grids using the node method according to [BBR95]. As this and other work [Gro12], this thesis relies on the following assumptions:

- incompressible transport medium

- no leakages
- perfect mixing in nodes (one single outflow temperature per node)
- heat propagates only with the mass flow (advection), no conduction, no diffusion
- return temperatures are constant
- pressure dynamics are not considered as they are too fast
- specific heat capacity and density of water are constant in the relevant temperature range (50 °C to 130 °C)
- thermal demand of consumers is known and given for each time slot

The key concept of the node method is to track the propagation of a volume of water at a certain temperature in the pipe. A district heating grid is described by nodes and pipes connecting the nodes. Thermal energy can be injected or withdrawn from the grid at nodes, if thermal generation sites or thermal loads are present. Besides generator parameters and load profiles, additional technical information like heat capacities and pipe diameters is required. The node method is a discrete time approach calculating temperatures at and mass flows between the nodes. Based on past temperatures of nodes and past mass flows between the nodes, the node temperatures for the next time step are calculated. Temperatures at intermediate points between the nodes are not calculated.

With the aforementioned assumptions (incompressible transport medium and no leakages) and assuming no storage tanks in nodes, there is a balance of mass flows $\dot{m}_{i,t}$ flowing into and out of a node:

$$\sum_{i \in inflow} \dot{m}_{i,t} = \sum_{i \in outflow} \dot{m}_{i,t} \qquad \forall t \in S_t. \tag{2.7}$$

Similarly, the mass flow into and out of a pipe are equal.

Assuming perfect mixing of inflow temperatures $T_{i,t}^{in}$ results in a single outflow temperature T_t^{out} for a node, which can be calculated as

$$T_t^{out} = \frac{\sum_{i \in inflow} \left(T_{i,t}^{in} \cdot \dot{m}_{i,t} \right)}{\sum_{i \in inflow} \dot{m}_{i,t}} \qquad \forall t \in S_t. \tag{2.8}$$

Thermal generation injecting heat into the system consumes water from a node of the return pipe network, heats it and feeds it to a node of the supply pipe network. Thermal loads withdrawing heat from the system consume water from a node in the supply pipe network, chill it and feed it to a node of the return pipe network. Heating and chilling of water leads to a change of its inner energy Q_t which can be described by

$$Q_t = c_p \dot{m}_t \left(T_t^{supply} - T_t^{return} \right) \qquad \forall t \in S_t \tag{2.9}$$

where $\dot{m}_t$ is the mass flow between supply and return side and c_p is the specific heat capacity of the transport medium water (assumed to be constant). T_t^{supply} and T_t^{return} are the temperatures at supply and return side.

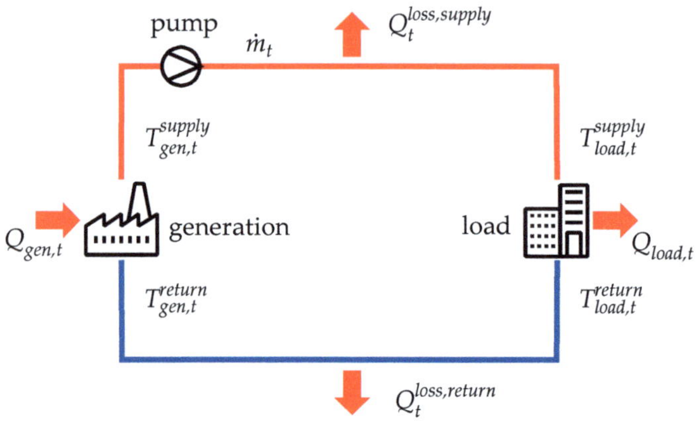

Figure 2.2: A heating grid with one generation and one load. Red lines: supply pipes; blue lines: return pipes; red arrows: thermal energy injection, retrieval and losses.

Figure 2.2 shows the most simple setup for a heating network with one thermal generation asset and one thermal load. Thermal losses occur in both supply and return network. However, thermal losses in the supply network are more important than in the return network due to the larger temperature difference to the surrounding area. Thermal loads usually have a controller adjusting the mass flow from supply to return side such that a given return temperature is reached for the arriving supply temperature and the current demand. As there is a pump close to the generation site with a constant pressure output, a valve at the load is sufficient for this mass flow control. The supply temperature is controlled at thermal generation sites. The return temperature depends on the return temperature arriving from the consumers. For all but one generation unit, the mass flow is controlled to fit the scheduled thermal energy supply of this generation unit. For the remaining generation unit, the mass flow and heat output remain variable to cope with unforeseen changes in thermal demand.

With this very common control strategy, the operators of a heating grid can control the temperature levels only in the supply lines directly. The temperature level in return pipes mainly depends on the controllers at the consumers whose behavior is hard to predict. The ability to control the temperature is crucial for storing thermal energy in a heating grid [Gro12]. Therefore, only the thermal dynamics in the supply lines, which are under direct control of the heating grid operator, are considered in this thesis and a constant return temperature is assumed.

As pressure dynamics travel with speed of sound in water (about 1200 m/s), they are a lot faster than the temperature dynamics which travel with flow velocity (at most 1.5 m/s to 2 m/s). Hence, pressure dynamics are neglected and only temperature dynamics are explained in the following.

In the node method the outflow temperature T_t^{out} of a pipe is calculated with a linear combination of past and present inflow temperatures T_t^{in}. Example 2.1 shows how this linear combination represents the propagation of temperature in a pipe.

Example 2.1:

Figure 2.3 shows the temperature distribution inside a pipe of length L at the beginning of a time slot t.

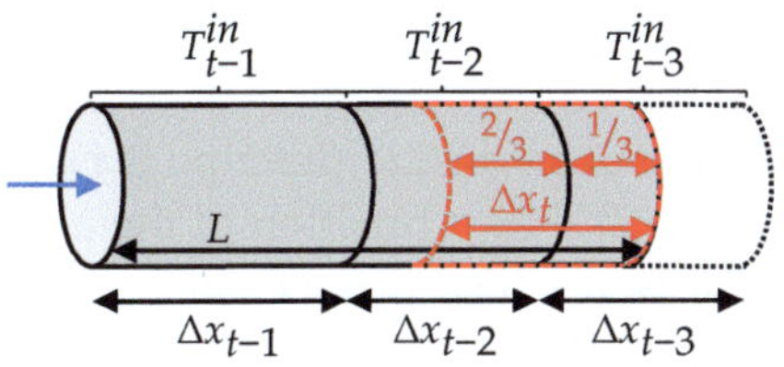

Figure 2.3: Scheme of pipe filled with different temperatures at beginning of a time slot t. The red-dashed volume leaves the pipe in time slot t. The blue arrow indicates the flow direction (left to right).

T_{t-1}^{in}, T_{t-2}^{in} and T_{t-3}^{in} are the inflow temperatures feeding the pipe in past time slots $t-1$, $t-2$ and $t-3$. Δx_t, Δx_{t-1}, Δx_{t-2} and Δx_{t-3} are the distance traveled by the water inside the pipe in time slots t, $t-1$, $t-2$ and $t-3$, respectively.
As in time slot t the water moves by Δx_t, the red-dashed volume in Figure 2.3 leaves the pipe in time slot t. Hence, the average outflow temperature of this time slot T_t^{out} is the average temperature of the red-dashed volume. If heat losses are not considered, the example shown in Figure 2.3 leads to (2.10) mixing the two temperatures inside the red-dashed volume with a weighted average:

$$T_t^{out} = \frac{2}{3}T_{t-2}^{in} + \frac{1}{3}T_{t-3}^{in}. \tag{2.10}$$

The propagation of the transport medium in the pipe Δx_t in time slot t can be calculated using velocity v_t or mass flow $\dot{m}_t$ with Δt being the duration of a time slot, ρ being the density of the transport medium (water) and d being the diameter of the pipe. The resulting equation is

$$\Delta x_t = \Delta t \cdot v_t = \frac{\Delta t}{\rho \pi \left(\frac{d}{2}\right)^2}\dot{m}_t \qquad \forall t \in S_t. \tag{2.11}$$

Following Example 2.1, the pipe outflow temperature T_t^{out} in every time slot t can be calculated as a linear combination of past inflow temperatures T_t^{in} with maximum transport delay τ^{max} and weight factors $\alpha_{\tau,t}$:

$$T_t^{out} = \sum_{\tau=0}^{\tau^{max}} \alpha_{\tau,t} T_{t-\tau}^{in} \qquad \forall t \in S_t. \tag{2.12}$$

To consider heat losses, let us first have a look at the conductive losses for a pipe filled with a liquid at one temperature T^{inside}. Here, the heat losses q^{loss} can be described with

$$q^{loss} = k_V \pi d L \left(T^{inside} - T^{amb}\right) \tag{2.13}$$

using diameter d and length L of the pipe, ambient temperature T^{amb} and heat loss factor k_V. This loss factor can be found in the pipe data sheet or be calculated based on material and construction.

In a dynamic setup the temperature inside the pipe is not constant. Thus, to calculate the pipe outflow temperature T_t^{out} as a function of the ambient temperature T^{amb} and past pipe inflow temperatures, the heat (loss) balance for one infinitesimal volume element with length dx traveling through the pipe at position x needs to be calculated. Now let us combine (2.9) and (2.13) and adapt both for an infinitesimal volume element $\mathrm{d}x$. With $T(x)$ as the temperature of the infinitesimal volume element at position x and $\mathrm{d}T(x)$ as the change of this temperature, we get

$$\pi d k_V \mathrm{d}x \left(T(x) - T^{amb}\right) = \dot{m} c_p \mathrm{d}T(x). \tag{2.14}$$

Solving this in-homogeneous differential equation for $T(x)$ leads to

$$T_t^{out} = T^{amb} + \left(T_{t-\tau_t}^{in} - T^{amb}\right) e^{-\frac{4 \cdot k_V}{\rho c_p d} \cdot \tau_t} \tag{2.15}$$

for the pipe outflow temperature T_t^{out} depending on ambient temperature T^{amb}, the travel time needed to pass the pipe τ_t, the past inflow temperature $T_{t-\tau_t}^{in}$ and pipe parameters like heat loss coefficient k_V and diameter d.

In a discrete time setting, the travel time τ_t can be calculated based on the number of time slots needed to pass the pipe. Using a simulation environment supporting continuous time equations like Modelica, (2.15) can be implemented directly in combination with

$$\tau_{t+1} - \tau_t = 1 - \frac{v_t}{v_{t-\tau_t}} \tag{2.16}$$

calculating the travel time τ_t based on the velocity v_t [SFT$^+$05, Gir16].

Extensions of the node method

An extension of the node method to a continuous time model is presented in [SFT$^+$05, Gir16] which enables exact formulations in the model and simulation environment

Modelica. Oppelt presents a refinement of the discrete time node method in [Opp15], which tracks the propagation of temperature not only across one but across several pipes. Thus, averaging of temperatures occurs less often and results get more accurate. However, this refinement of the node method addresses simulation and is not straight forward to use in optimization of thermal grids.

2.1.3 Storing thermal energy in a heating grid

If the heating grid dynamics introduced in the previous section are used in a smart way, thermal energy can be stored in a heating grid. In Example 2.2, this storage effect is explained for a simple heating grid with one producer and one load. This example follows our publication [MHH19].

Example 2.2:

We consider the most simple setup of a heating grid with one producer and one load connected with a supply and a return pipe as shown in Figure 2.4. A supply temperature profile at the producer is given with an increase of temperature in time step 3 and a decrease of temperature in time step 8. Perfect control of return temperature is assumed at the consumer leading to a constant return temperature. In average the supply temperature is slightly lower at the consumer due to thermal losses in the pipe. A constant thermal demand is assumed for the full time horizon.

As the flow velocity in the pipe is limited, the temperature increase reaches the consumer with a transport delay. Hence, despite the temperature increase in time step 3 the mass flow first remains unchanged which leads to an immediate increase of thermal generation of the producer according to (2.9). As soon as the increased supply temperature reaches the consumer after two time steps in time step 5, the controller at the consumer reduces the mass flow to keep (2.9) in balance. As hydraulic dynamics are very fast (see Section 2.1.2) the mass flow is reduced almost immediately at the producer, leading to a reduction of thermal energy output in time step 5. Thus, in time step 3 and time step 4, when the increase of temperature did not reach the consumer yet, the thermal energy produced at the producer exceeds the thermal demand of the consumer and the thermal losses. Hence, here the thermal dynamics of the heating grid show a behavior similar to charging an energy storage.

Reducing the supply temperature at the producer in time step 8 leads to an inverse behavior. As the decrease of supply temperature is not seen at the consumer before time step 10, the mass flow stays constant until time step 10 and, thus, the thermal energy output of the producer is reduced proportionally to the supply temperature decrease. As soon as the supply temperature decrease reaches the consumer, its controller will increase the mass flow in order to supply the constant heat load. With this additional mass flow, the thermal energy output of the producer is increased back to its original level. In the period of time step 8 and 9, the thermal energy output of the producer is below the thermal load plus thermal losses. Thus, it shows a similar behavior as a discharge of an energy storage. Hence, the thermal dynamics of a heating grid show an energy-storage-like behavior where increases of supply temperature lead to charging of the storage and decreases of supply temperature lead to discharging of the storage.

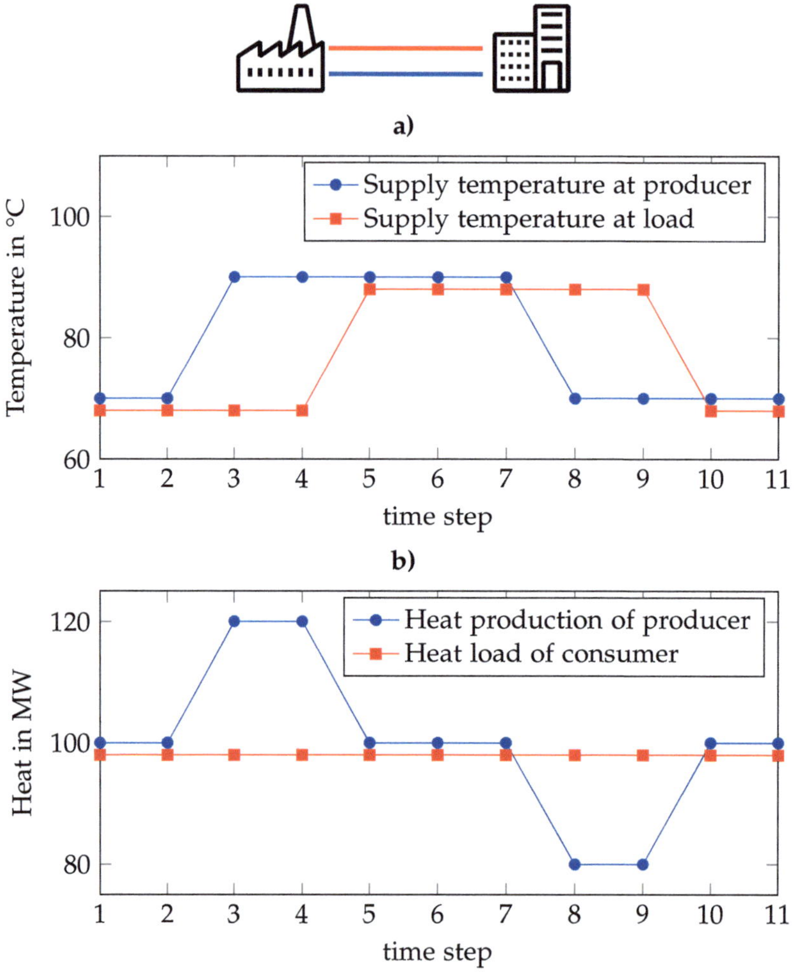

Figure 2.4: Storage behavior of a heating grid with one CHP and one consumer. a) Supply temperatures at producer and load. b) Heat production of producer and heat consumption of consumer.

Remark: *As the mass flow is reduced in times of high supply temperature at the consumer, in reality the transport delay increases. Thus, the decrease of temperature in time step 8 would arrive slightly later at the consumer. For high temperature differences this effect is more important than for small temperature changes.*

2.1.4 Optimization of heating grids using the node method

To utilize the storage effect explained above, several approaches in literature have been proposed. Unfortunately, the mass flow dependent time delay induced by the transport time in (2.15) as well as the bilinear terms in (2.9) lead to a non-convex problem. Thus, finding optimal solutions is very complex.

A common approach used in many publications is to fix the mass flows or time delays prior to optimization, as this leads to a linear and thus convex problem. Some publications assume directly that the mass flows in the heating grid are known upfront. This is the case when they are operated with a different control strategy without valves at the consumer side and, therefore, no changes in mass flow are induced by the consumers [LWW+16, GWL+17, LWLL17, ZFZC18, ZZZW18]. Unfortunately, assuming that consumers do not control the flow means that no thermostats are used, which does not allow accurate local control of the indoor temperature. Hence, assuming constant mass flow can be a valid assumption in some areas, but for most heating grids worldwide this assumption is not valid.

To overcome this issue, several researchers propose to use an iterative or sequential solution approach. The authors of [BBR95] use a model with fixed time delays in iteration with a model with fixed supply temperatures to optimize operations of heating grids aiming for a fully dynamic optimization. Another iterative scheme is used in [SFT+05]. Here parameters of an optimization model with fixed time delays are iteratively updated using a simulation model. Robustness of the solution is increased by requiring an additional heat injection. A similar iterative approach is used in [Gir16, GMBV17] where results of a detailed simulation update the parameters of a linear grid model. To get a linear model, fixed time delays from generation to consumers are assumed for optimization. In [DCM+19] the authors as well use an iterative optimization scheme where an optimization model with fixed delay times is updated using a simulation of the heating grid. In addition, the thermal inertia of buildings is considered as thermal storage opportunity .

An intra-day optimization approach using non-linear optimization models is proposed in [LWS+16]. As intra-day approach it does not decide on generation unit commitment and does not consider forecast uncertainty. To cope with transport delays it uses an approach different from the papers mentioned before: Several complicating variables are identified for a pipe model based on the node method (including integer time delays). Those complicating variables are fixed in a non-linear optimization model which is updated using a simulation model in every iteration. Similarly, the approach proposed in [Trö99] does not use fixed time delays. It relies on the principle ideas of the node method for deriving suitable models, but uses an iterative approach combining a static hydraulic, a static thermal and a dynamic thermal model to optimize the dynamics of a heating grid.

2.1.5 Optimization of heating grids: other approaches

Despite many publications using the node method to model the thermal dynamics of a heating grid, there are a some papers using different approaches. Starting from a general thermal energy balance a PID control approach considering the inherent storage capacity of a heating grid is presented in [BJPS11]. No optimization, but a parameter study is used to identify good times for charging or discharging the district heating grid's storage capabilities. Lesko and Bujalski propose to model thermal dynamics of heating grids using a simple energy balance [LB17]. Here, the heating grid is assumed to be one single water volume with perfect mixing. Increasing or decreasing the supply temperature charges or discharges the thermal capacity of this volume. This modeling approach is compared with a model derived by the node method and parameterized for real-world data. It is shown that accuracy of both modeling approaches depends on a good parameterization [LB17].

In [Gro12], a rather different approach is used: The heat storage capabilities of a radial heating grid are approximated with a linear regression model which is trained based on a high number of heating grid simulations. Those grid simulations require a detailed physical model of the grid, whereas the resulting linear regression model is a lightweight model linking the thermal energy stored inside the grid with past and current supply temperatures and the current load factor of the grid. Hence, it enables fast optimization runs as the computational effort is moved from optimization to model parameterization.

In [VD15], a detailed simulation model built in MATLAB/Simulink derives heat demands and mass flows in a meshed heating grid. Those results are used as inputs for an optimization of the generation sites. This work is extended in [VTD17] with a hybrid-evolutionary algorithm combining a MILP at lower level and a genetic algorithm at upper level to minimize total operating cost of a meshed heating grid. However, thermal dynamics of heating grids are only considered in steady state in the simulation model run by the genetic algorithm.

A decomposition of the mixed-integer non-linear problem (MINLP)[2] of district heating optimization into a unit commitment and an economic dispatch problem is proposed in [SLM+17]. The unit commitment problem is formulated as mixed integer quadratic constraint program without consideration of the heating grid dynamics leading to a problem similar the one presented in Section 2.1.1 except having a quadratic instead of a linear cost function. The economic dispatch problem is a non-linear program (NLP)[3] which is formulated in the object-oriented modeling language

2 Like MILPs, MINLPs are optimization problems consisting of real and integer variables. However, for MINLPs non-linear objective functions as well as non-linear equality and inequality constraints are possible. As MINLPs are hard to solve, off-the-shelf solvers only exist for special types of non-linearity.

3 In difference to MINLPs, NLPs are optimization problems only consisting of real variables. Off-the-shelf solvers finding local optima are available for NLPs.

Modelica and considers full dynamics of the heating grid. A local solver is used to solve this NLP.

As methods for optimization of district heating grids are usually computationally expensive, there have been efforts to speed up simulation and optimization of heating grids. A Danish and a German method for aggregation of the grid topology trying to loose as little accuracy as possible have been proposed. A comparison of the two is published in [LBW04, WBS05].

Considering different forms of energy beyond district heating, Geidl and Andersson propose a joint optimal power flow for electricity, heat and gas networks [GA07]. To enable applicability to large scale systems, a decomposition scheme for the aforementioned approach is presented in [MAAFFH14]. However, for both publications heating grid dynamics are not in focus and variable transport delays in temperature propagation are not considered. In [LM17, ZFZC18], a combination of electricity, gas and heat networks is considered as well, whereas the building thermal inertia is used as additional thermal energy storage besides the heating grid thermal inertia in [LWLL17, GWL+17].

A completely different storage concept is proposed in [DMOK19]. Here, instead of using the thermal dynamics induced by variations of supply temperature, the flow direction is reversed in order to store thermal energy inside a heating grid. This approach for thermal storage is only applicable if large thermal generation exists at both sides of a pipe used for storage. Hence, it implies strong requirements on system setup and grid topology and, therefore, is not considered further in this thesis.

2.2 Global optimization of non-convex problems

Global optimization of non-convex problems is a complex topic studied for many decades [HT96]. A common, well-established approach proving global optimality of a solution, is to transform the non-convex problem into a primal and a dual problem which both are easier to solve than the original non-convex problem [FV93]. The primal problem leads to a feasible solution of the original problem, for example by searching for a solution only inside a reduced solution space. For a minimization problem, this solution of the primal problem represents an upper bound of the original problem, as better solutions might be possible in the original solution space. The dual problem is an outer-approximation of the original problem. Very often it is chosen to be convex in order to ensure global optimality of its solution. It increases the original solution space such that all feasible solutions of the original problem are part of the solution space of the dual problem, but additional solutions are possible. Hence, for minimization problems, the solution of the dual problem represents a lower bound of the original problem. Solutions of the original problem cannot be better, but the dual solution might not be a feasible solution of the original problem. To solve the original non-convex problem, the primal and dual problem are solved in multiple iterations. If primal and dual solution are equal, the globally optimal solution is reached. If they are not equal, the primal solution is a feasible solution of the original problem and the dual solution indicates the gap to global optimality. If in every iteration the exactness of the dual and primal solutions is improved, this gap is reduced. The iterations usually stop, when a specific numeric precision of the solution is reached (e.g. a gap smaller than 1 %) [FV93].

For some non-convex problem classes like MILPs, solvers using such a primal-dual algorithm reaching proven global optimality are available as open source [BEN05, FRV+18] or commercial tools [Gur16].

In the following, only the problem classes relevant for modeling and optimization of district heating grids are discussed where no off-the-shelf solvers exist. These are bilinear terms like in (2.9) or (2.10) and a variable-dependent time delay as in (2.15). Several approaches exist for global optimization of bilinear problems as discussed in Section 2.2.1 and Section 2.2.2. Necessary conditions for a local optimum of variable-dependent time delays are defined in [CGCP16] and a solution algorithm finding such local optima is proposed in [CPB17]. However, for problems with variable-dependent time delays, no approach proving global optimality of its solution has been presented yet. Global optimization solvers for NLPs or MINLPs like BARON [Sah96] do also not support solving variable-dependent time delays.

2.2.1 McCormick envelopes

McCormick envelopes are commonly used as outer-approximation or dual problem for global optimization of bilinear terms with bounded multipliers, because they are the tightest convex hull of a bilinear term with bound variables [McC76]. Figure 2.5 shows the product of two variables, each bounded by 1 and 5, with under-estimating McCormick envelopes. It can easily be seen that combining the planes in the two sub-figures provides a tight convex hull of the bilinear term. In combination with a branching strategy or a piece-wise linear approximation, McCormick envelopes can be used to build a global optimization scheme for bilinear terms.

Therefore, McCormick envelopes are employed in several approaches for global optimization of bilinear terms. In [AKF83], a branch and bound solution algorithm using McCormick envelopes as over- and under-estimators of a bilinear function is presented. A piece-wise McCormick relaxation is used in [dACZ+17] for handling bilinear terms in a scheduling problem for operation of crude oil terminals. A piece-wise linear relaxation with McCormick envelopes using bivariate partitioning is presented in [HK10] to solve bilinear programs.

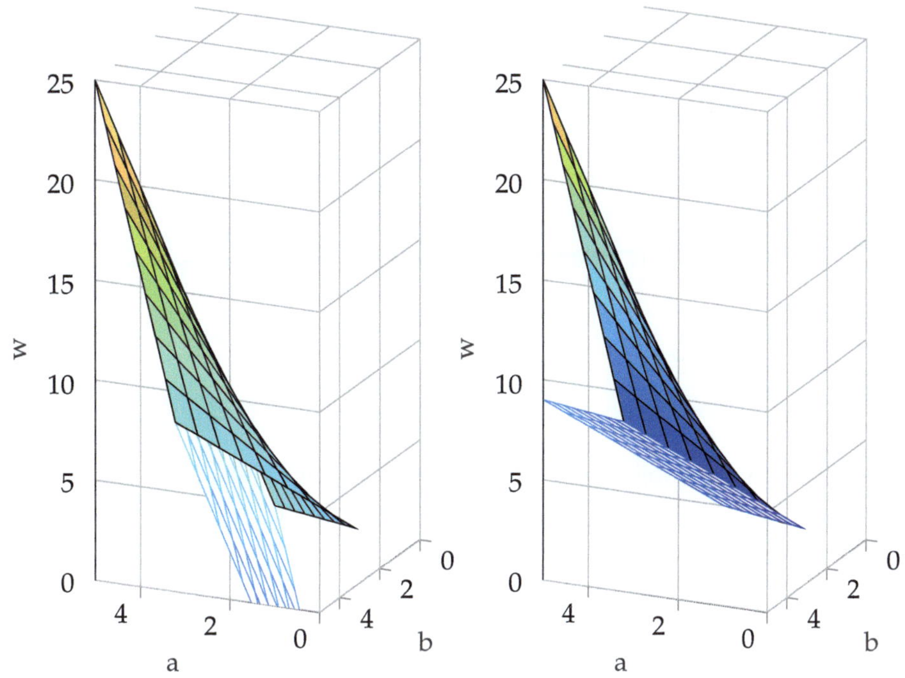

Figure 2.5: Surface plots of product $w = a \cdot b$ with $a, b \in [1,5]$ and planes of the McCormick under-estimators

2.2.2 Multiparametric disaggregation

In this thesis, multiparametric disaggregation [TCM12] is used for global optimization of bilinear terms as it promises faster solution times than a piece-wise linear approximation using McCormick envelopes [CT13].

The concept of multiparametric disaggregation is the discretization of one of the variables in a product $w_{i,j} = x_i \cdot x_j$. Like with floating point numbers in computing, every digit of this variable is split into multiple binary variables which encode the value of this digit. Here, variable x_j is split into binary variables z_{jkl} with l representing the position of the digit and k taking values from 0 to 9 representing the value of this digit. Hence, as every digit can take only one value, z_{jkl} is non-zero only for one k for all l and j [TCM12].

$$x_j = \sum_{l=p}^{P} \sum_{k=0}^{9} 10^l \cdot k \cdot z_{jkl} + \Delta x_j \tag{2.17}$$

$$\sum_{k=0}^{9} z_{jkl} = 1 \qquad \{\forall l \in \mathbb{Z} | p \leq l \leq P\} \tag{2.18}$$

As l defines the position of the digit, p and P define the precision of approximation. If p approaches negative infinity and P is large enough, (2.17) becomes exact. Δx_j is a slack variable bounded by $0 \leq \Delta x_j \leq 10^p$ which transforms the product $w_{i,j}$ according to [TCM12] into

$$w_{i,j} = \sum_{l=p}^{P} \sum_{k=0}^{9} 10^l \cdot k \cdot \hat{x}_{ijkl} + \Delta w_{ij}. \tag{2.19}$$

Here $\hat{x}_{ijkl} = x_i \cdot z_{jkl}$ which is enforced with

$$x_i = \sum_{k=0}^{9} \hat{x}_{ijkl} \qquad \{\forall l \in \mathbb{Z} | p \leq l \leq P\} \tag{2.20}$$

$$x_i^L \cdot z_{jkl} \leq \hat{x}_{ijkl} \leq x_i^U \cdot z_{jkl} \qquad \{\forall k \in \mathbb{Z} | 0 \leq k \leq 9\}, \{\forall l \in \mathbb{Z} | p \leq l \leq P\} \tag{2.21}$$

where x_i^L and x_i^U are the lower and upper bounds of variable x_i [TCM12].

The slack variable Δw_{ij} is linked to the slack variable Δx_j by $\Delta w_{ij} = x_i \cdot \Delta x_j$ and accordingly [TCM12]

$$x_i^L \cdot \Delta x_j \leq \Delta\, w_{ij} \leq x_i^U \cdot \Delta x_j. \tag{2.22}$$

Due to the slack variable Δw_{ij} the solution space of (2.19) includes the original solution of the product $w_{i,j} = x_i \cdot x_j$ and, thus, it is an outer approximation [TCM12]. Hence, this yields a MILP which is a dual (or lower bound) problem of the original problem. The iterative scheme shown in Figure 2.6 is used to solve the bilinear problem to global optimality, using the non-linear model as primal (or upper bound) problem being solved with a local, non-linear solver [TCM12].

For ease of understanding, multiparametric disaggregation is introduced using a base of 10 ($k = 0$ to 9) in this section. Choosing a different base is possible potentially leading to faster solution times [TCM12].

2.3 Pipeline scheduling: modeling of flow dynamics in other areas

In addition to the optimization approaches presented in Sections 2.1.4 and 2.1.5, optimization models for related problems can inspire innovative solutions for heating grid optimization. In this section, pipeline scheduling is introduced as it is a well studied field of research with many similarities to heating grid optimization. In pipeline scheduling, liquid refined petroleum products need to be transported via a pipeline system from refineries to depots. There are multiple products, but only a limited number of pipes connecting refineries and depots.

In Figure 2.7 a pipeline scheduling example presented in [MC17] is shown. There are three pipe segments in this example. Pipe segment S1 connects the input node with

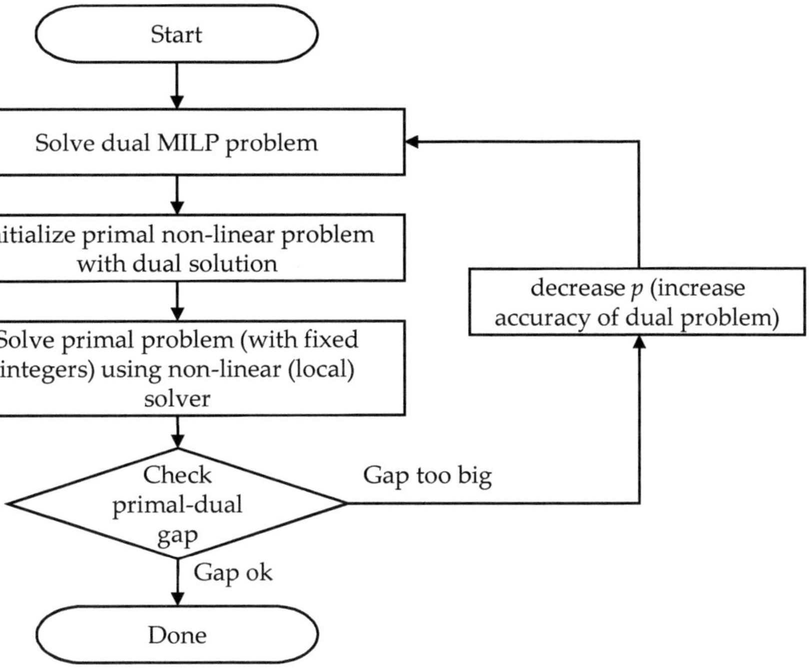

Figure 2.6: Iterative solution approach of multiparametric disaggregation

refinery R1 to the output node with depot D1. Segment S2 connects the output node with Depot D1 to the dual purpose node DP1 with input by refinery R2 and output to depot D2. Pipe segment S3 connects dual purpose node DP1 to the output node with depot D3. There are five different products P1 to P5 which can be produced at refineries R1 and R2. Depots D1, D2 and D3 each require a certain amount of these products. The goal of optimal pipeline scheduling is to find an optimal product sequence transported in the pipes allowing to achieve all demands at minimum cost or in minimum time. Pipeline scheduling for refinery products is having many similarities with optimization of heating grids: There are nodes feeding into the pipeline system and nodes consuming from the pipeline system. The dynamics of propagation of a liquid in a pipe are similar, too. The liquid propagates with the mass flow and there is a transport time from entering to leaving the pipe. However, in pipeline scheduling for refinery products there is a discrete set of different products which is transported via the pipes, whereas in heating grids the supply temperature is a real variable which can be chosen freely within its limits.

In [RP04], a discrete volume representation of the pipeline in combination with a discrete time grid for the scheduling horizon is introduced to optimize the pipeline scheduling problem for refinery products. However, using a continuous time formulation of flow dynamics, [CC04] allows a MILP formulation. Today's models for pipeline scheduling are based on this concept and can handle reversible mass flows

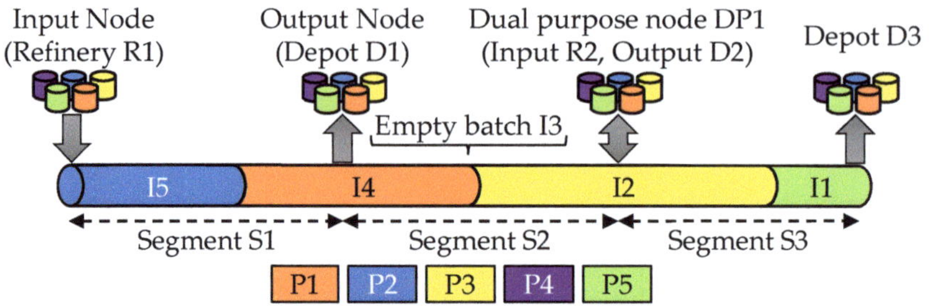

Figure 2.7: Pipeline distribution system for refined petroleum products after [MC17].

and different network topologies using a continuous time formulation as well as a continuous representation of volumes [Cas17]. A major difference of this product-centric model formulation to other model concepts like [RP04] is that the variable flow rates are considered only implicitly without being an explicit optimization variable. This allows a very efficient MILP continuous time formulation [CGZ18] avoiding non-linearities in the optimization. However, it prevents an accurate modeling of pumping cost [CCMC15].

2.4 Objective of this thesis

CHPs need to operate more and more flexible to react to volatile renewable generation, volatile demand and volatile energy prices which requires thermal storage. The thermal dynamics of a heating grid allow to store thermal energy. Thus, they enable a flexible operation of CHPs without requiring an investment in dedicated thermal storage tanks. To leverage this potential, the thermal dynamics of a heating grid need to be considered in operations planning. Model formulations for these dynamics usually include bilinear terms and a variable-dependent time delay representing the temperature propagation in a pipe. Thus, they result in non-convex problems and the approaches proposed to solve these problems do not prove the global optimality of their solution as shown in Section 2.1.4 and Section 2.1.5. In the literature on global optimization, there are many approaches for bilinear problems (see Section 2.2.1 and Section 2.2.2). However, there are no approaches for global optimization of problems with variable-dependent time delays.

Thus, the approaches for heating grid optimization presented in the literature do not offer a guarantee to reach the global optimum and the literature on global optimization does not offer appropriate solution algorithms either. Nevertheless, it is essential to find a solution as close as possible to the global optimum to operate a heating grid as sustainable as possible and to leverage the full potential of using the thermal dynamics of a heating grid as energy storage. Accordingly, this thesis aims at finding the globally optimal solution for scheduling of CHPs considering heating grid dynamics.

In addition, current optimization approaches considering the thermal dynamics of heating grids are only very rarely applied in real world installations. Thus, this thesis aims as well at finding real world applicable optimization models which enable fast and reliable solutions. A model training using historic measurements would be ideal to enable easy brown field installations without a major model parameterization effort.

In summary, the following goals in modeling and optimization of heating grids will be addressed in this thesis:

- Find a globally optimal solution for optimization of heating grids allowing to leverage the full potential of considering the thermal dynamics in operations planning
- Find fast and reliable approaches for optimization of heating grids with accurate solutions and easy model parameterization enabling real world applications

3 Global optimization with multiparametric delay modeling

Existing approaches as discussed in Section 2.1 do not prove global optimality of their solutions as they simplify the non-convex heating grid dynamics in their optimization models. Hence, it is not possible to judge the quality of their solution as it is unknown if a better solution exists. Non-convexities in heating grid dynamics are bilinear terms in (2.8) and (2.9) as well as a variable-dependent time delay in (2.15). Global optimization of bilinear terms is a well studied field as discussed in Section 2.2. However, there are no methods for global optimization of problems with variable-dependent time delays. In the following, "multiparametric delay modeling" is presented, that is able to deal with this problem using a primal-dual algorithm as discussed in Section 2.2. Starting with the dual problem in Section 3.1, this chapter outlines two methods to formulate the primal problem in Section 3.2 and proposes the resulting primal-dual algorithm in Section 3.3.

"Multiparametric delay modeling" is the first approach for global optimization of systems with variable-dependent time delays. It enables to globally optimize the fully dynamic problem of optimal operation of heating grids if combined with a global optimization approach for bilinear terms like multiparametric disaggregation presented in Section 2.2.2. The basic idea of "multiparametric delay modeling" was first published in [MH19]. More details are published in [MLH19] with a discussion of the primal and dual problems and a presentation of proofs of convergence to the global optimum. This chapter presents the enhanced model formulation as in [MLH19] as it improved significantly solution quality and solution times.

The optimization model for global optimization of heating grids is based on the unit commitment problem without thermal dynamics (Section 2.1.1) where the heat balance equation (2.5) is replaced with the dynamics of the node method (Section 2.1.2). The following sections discuss how to model the non-convexities of this problem formulation in the primal and the dual problem.

3.1 Dual problem

As a first step to a primal-dual global optimization algorithm, this section introduces the outer approximation or dual problem of variable-dependent delay times based on the node method (Section 2.1.2) as in [MLH19]. The basic idea of this outer approximation of the pipe outflow temperature with "multiparametric delay modeling" is explained in Example 3.1.

Example 3.1:

As in Example 2.1, Figure 3.1 shows a temperature distribution inside a pipe at beginning of time slot t. The pipe of length L is filled with volumes at temperature levels of past inflow temperatures T^{in}_{t-1}, T^{in}_{t-2} and T^{in}_{t-3}. Those volumes span a width of Δx_{t-1}, Δx_{t-2} and Δx_{t-3} being the propagation of water inside the pipe in the previous time slots $t-1$, $t-2$ and $t-3$.

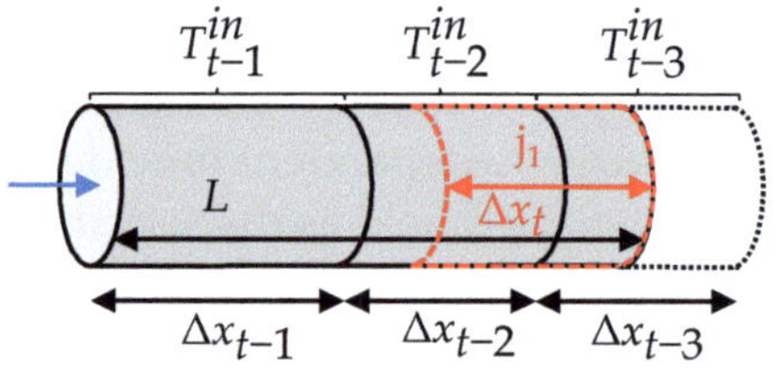

Figure 3.1: Pipe filled with different temperatures at beginning of time slot t. The red-dashed volume j_1 leaves the pipe in time slot t. The blue arrow indicates the flow direction (left to right).

The red-dashed volume j_1 in Figure 3.1 will leave the pipe in time slot t. Thus, its average temperature is the average pipe outflow temperature in this time slot. The temperature of this red-dashed volume j_1 is a mixture of volumes being at temperatures T^{in}_{t-2} and T^{in}_{t-3}. Hence, the average outflow temperature T^{out}_t in time slot t needs to be between those two temperatures. Thus, the following inequality leads to an outer approximation of the outflow temperature T^{out}_t in time slot t.

$$\min\left(T^{in}_{t-2}, T^{in}_{t-3}\right) \leq T^{out}_t \leq \max\left(T^{in}_{t-2}, T^{in}_{t-3}\right) \tag{3.1}$$

To refine this outer approximation, the volume leaving the pipe in time slot t is divided into two sub-volumes j_1 and j_2 of identical size in Figure 3.2.

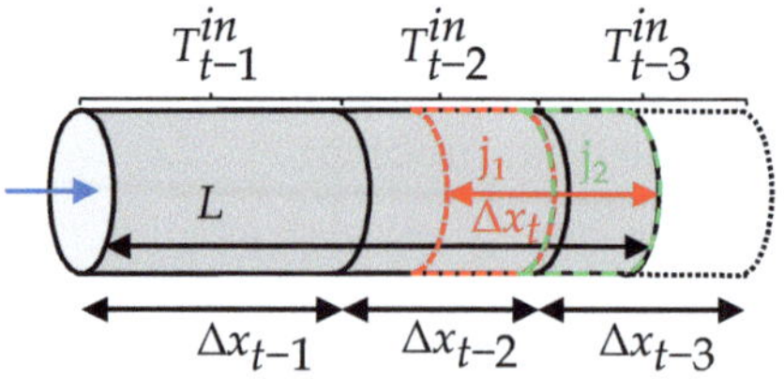

Figure 3.2: Pipe filled with different temperatures at beginning of time slot t. The volumes j_1 (red) and j_2 (green) leave the pipe in time slot t. The blue arrow indicates the flow direction (left to right).

Now the outflow temperature of sub-volume j_1 (red) T^{out}_{t,j_1} is known exactly as it only consists of water at temperature T^{in}_{t-2}. The outflow temperature of sub-volume j_2 (green) T^{out}_{t,j_2} can be estimated using a similar outer approximation as before for volume j_1 in Figure 3.1:

$$T^{out}_{t,j_1} = T^{in}_{t-2}, \tag{3.2}$$

$$\min\left(T^{in}_{t-2}, T^{in}_{t-3}\right) \leq T^{out}_{t,j_2} \leq \max\left(T^{in}_{t-2}, T^{in}_{t-3}\right). \tag{3.3}$$

To get the pipe outflow temperature T^{out}_{t} in time slot t assuming perfect mixing of sub-volumes j_1 and j_2, the average of the temperatures of the sub-volumes needs to be calculated with

$$T^{out}_{t} = \frac{1}{2}T^{out}_{t,j_1} + \frac{1}{2}T^{out}_{t,j_2}. \tag{3.4}$$

All in all, for the case shown in Figure 3.2 we get

$$\frac{1}{2}T^{in}_{t-2} + \frac{1}{2}\min\left(T^{in}_{t-2}, T^{in}_{t-3}\right) \leq T^{out}_{t} \leq \frac{1}{2}T^{in}_{t-2} + \frac{1}{2}\max\left(T^{in}_{t-2}, T^{in}_{t-3}\right) \tag{3.5}$$

as outer approximation for the pipe outflow temperature T^{out}_{t} in time slot t.
In (3.5) the temperature of only half of the volume leaving the pipe in time slot t is estimated. Hence, (3.5) is a tighter outer approximation than the outer approximation in (3.1) and adding a sub-volume increased tightness of the outer approximation.

Example 3.1 shows a way to formulate outer approximations for the outflow temperature of a pipe including a possibility to refine this outer approximation by introducing sub-volumes. Thus, if there is a way to formulate this approach in general as a solvable problem, it can be used as dual in a primal-dual global optimization scheme. The following presents a general MILP formulation of this concept as in [MH19, MLH19].

The non-negative real variable Δx_t represents the distance traveled by the transport medium inside the pipe in one time slot t. It is calculated using (2.11):

$$\Delta x_t = \Delta t \cdot v_t = \frac{\Delta t}{\rho\pi\left(\frac{d}{2}\right)^2}\dot{m}_t \qquad \forall t \in S_t. \tag{3.6}$$

Based on this distance traveled Δx_t, we introduce the binary variable $b_{\tau,j,t}$ representing a discrete time delay. This binary variable $b_{\tau,j,t}$ shall equal 1 if and only if sub-volume j leaving the pipe in time slot t entered the pipe τ time slots ago. Hence, for every time slot t and every sub-volume j exactly one time delay τ exists and all other $b_{\tau,j,t}$ have to be zero, which can be enforced by

$$\sum_{\tau=0}^{t} b_{\tau,j,t} = 1 \qquad \forall t \in S_t, \forall j \in S_j. \tag{3.7}$$

Here, S_j denotes the set of sub-volumes j. This set is limited by the number of sub-volumes n_p: $S_j \in [1, n_p]$.

To ensure that the correct time delay τ is selected, the following inequalities are introduced. They compare the sum of past distances traveled within the pipe (Δx_t) with the length of the pipe L adjusted by the sizes of the j sub-volumes. Using a sufficiently large parameter $M^\star$ allows to set the correct discrete delay variable $b_{\tau,j,t}$ to 1 for every sub-volume j in every time slot t with

$$L - \frac{j-1}{n_p}\Delta x_t \leq \sum_{tt=t-\tau}^{t-1} \Delta x_{tt} + (1 - b_{\tau,j,t}) \cdot M^\star \qquad \forall t \in S_t, \forall j \in S_j, \forall \tau \in S_\tau, \tag{3.8}$$

$$L - \frac{j-1}{n_p}\Delta x_t \geq \sum_{tt=t-\tau+1}^{t-1} \Delta x_{tt} - (1 - b_{\tau,j,t}) \cdot M^\star \qquad \forall t \in S_t, \forall j \in S_j, \forall \tau \in S_\tau. \tag{3.9}$$

The set S_τ contains all possible (discrete) time delays.

Next, we need to find suitable outer approximations of the outflow temperatures of these sub-volumes $T^{out}_{t,j}$. Here several different cases need to be considered depending on the number of possible temperatures within a sub-volume.

First, if the temperature inside the sub-volume is constant, as for sub-volume j_1 in Figure 3.2, the outflow temperature of the sub-volume $T^{out}_{t,j}$ is known exactly. This is the case when the discrete time delays $b_{\tau,j,t}$ of two neighboring sub-volumes are equal. Thus, with T^U being the upper bound of the temperature we can assign the correct past inflow temperature $T^{in}_{t-\tau}$ to the outflow temperature of the sub-volume $T^{out}_{t,j}$ using

$$T^{out}_{t,j} \leq T^{in}_{t-\tau} + (2 - b_{\tau,j,t} - b_{\tau,j+1,t}) \cdot T^U \qquad \forall t \in S_t, \forall j \in S_j, \forall \tau \in S_\tau \tag{3.10}$$

$$T^{out}_{t,j} \geq T^{in}_{t-\tau} - (2 - b_{\tau,j,t} - b_{\tau,j+1,t}) \cdot T^U \qquad \forall t \in S_t, \forall j \in S_j, \forall \tau \in S_\tau. \tag{3.11}$$

Second, two different temperature levels are possible, if the discrete time delays $b_{\tau,j,t}$ of two neighboring sub-volumes j and $j+1$ are different by 1 ($b_{\tau,j,t}$ and $b_{\tau-1,j+1,t}$ are non-zero). Thus, an outer approximation of the outflow temperature as proposed in Example 3.1 can be used leading to

$$T^{out}_{t,j} \leq \max\left(T^{in}_{t-\tau}, T^{in}_{t-\tau+1}\right) + (2 - b_{\tau,j,t} - b_{\tau-1,j+1,t}) \cdot T^U \qquad \forall t \in S_t, \forall j \in S_j, \forall \tau \in S_\tau \tag{3.12}$$

$$T^{out}_{t,j} \geq \min\left(T^{in}_{t-\tau}, T^{in}_{t-\tau+1}\right) - (2 - b_{\tau,j,t} - b_{\tau-1,j+1,t}) \cdot T^U \qquad \forall t \in S_t, \forall j \in S_j, \forall \tau \in S_\tau. \tag{3.13}$$

A linear formulation allows an efficient solution of a MILP problem. Thus, the functions $\min(a,b)$ and $\max(a,b)$ need to be represented in linear equations. For this, we note that for any real, non-negative variables $a \in \mathbb{R}_{\geq 0}$ and $b \in \mathbb{R}_{\geq 0}$ the difference $a-b$ can be decomposed into two real, non-negative auxiliary variables $\delta^+ \in \mathbb{R}_{\geq 0}$ and $\delta^- \in \mathbb{R}_{\geq 0}$ as follows:

$$a - b = \delta^+ - \delta^- \tag{3.14}$$

Using these auxiliary variables we can enforce $c \leq \max(a,b)$ with

$$c \leq b + \delta^+. \tag{3.15}$$

Similarly for $c \geq \min(a,b)$ we get

$$c \geq b - \delta^-. \tag{3.16}$$

Third, if discrete delay times $b_{\tau,j,t}$ of two neighboring sub-volumes j and $j+1$ have a distance of 2 ($b_{\tau,j,t}$ and $b_{\tau-2,j+1,t}$ are selected), the sub-volume outflow temperature is a mix of three temperatures. Hence, the sub-volume outflow temperature needs to be between the minimum and the maximum of those three temperatures. In line with the outer approximation above we get

$$T_{t,j}^{out} \leq \max\left(T_{t-\tau}^{in}, T_{t-\tau+1}^{in}, T_{t-\tau+2}^{in}\right) + (2 - b_{\tau,j,t} - b_{\tau-2,j+1,t}) \cdot T^U \qquad \forall t \in S_t, \forall j \in S_j, \forall \tau \in S_\tau \tag{3.17}$$

$$T_{t,j}^{out} \geq \min\left(T_{t-\tau}^{in}, T_{t-\tau+1}^{in}, T_{t-\tau+2}^{in}\right) - (2 - b_{\tau,j,t} - b_{\tau-2,j+1,t}) \cdot T^U \qquad \forall t \in S_t, \forall j \in S_j, \forall \tau \in S_\tau. \tag{3.18}$$

The formulation from (3.14) to (3.16) for min and max functions can be extended to calculate the minimum or maximum of three variables using

$$\min(a,b,c) = \min(\min(a,b), c),$$
$$\max(a,b,c) = \max(\max(a,b), c).$$

The implementation of cases with more than three temperature levels in a sub-volume is not required. Indeed, by increasing the number of sub-volumes n_p we always obtain a situation with at most two temperatures per sub-volume as proven in Theorem 3.1. In early iterations of the optimization algorithm sub-volumes are still large. Thus, they easily span more than two or three temperature levels. To increase tightness of these early approximations, it is recommended to limit the sub-volume outflow temperature $\left(T^L \leq T_{t,j}^{out} \leq T^U\right)$ for all sub-volumes j and time slots t. In fact, the outer approximation for three temperature levels in (3.17) and (3.18) is theoretically not required to reach the global optimum as with a high number of sub-volumes this situation does not occur. However, it increases tightness of the formulation in early iterations, too, and this tighter formulation reduces solution time.

Theorem 3.1 (Decreasing number of temperatures per sub-volume)
There is an $n_p^* \in \mathbb{N}$ such that for all $n_p \geq n_p^*$ there are at most two temperatures inside any sub-volume $j \in S_j$.

Proof:
To prove Theorem 3.1, first note that the inflow temperature T_t^{in} into a pipe only changes at the beginning of a time slot t having a discrete time model. Thus, the distance between two neighboring temperatures T_{t-1}^{in} and T_t^{in} in a pipe is the progress of the transport medium within one time slot Δx_t which can be calculated according to (3.6).

With the proposed approach, the volume leaving the pipe in time slot t is divided into n_p equal-sized sub-volumes j. As the volume leaving the pipe in time slot t spans Δx_t, the n_p sub-volumes j leaving the pipe in time slot t span $\Delta x_{j,t}$:

$$\Delta x_{j,t} = \frac{\Delta x_t}{n_p} \qquad \forall j \in S_j, \forall t \in S_t. \tag{3.19}$$

If the number of sub-volumes n_p is increased, the span of one sub-volume $\Delta x_{j,t}$ decreases. Without any limitation on Δx_t we arrive at

$$\lim_{n_p \to \infty} \Delta x_{j,t} = \lim_{n_p \to \infty} \frac{\Delta x_t}{n_p} = 0 \qquad \forall j \in S_j, \forall t \in S_t. \tag{3.20}$$

Thus, there is always a number of sub-volumes n_p^* big enough to guarantee

$$\Delta x_{j,t} \leq \Delta x_{tt} \qquad \forall t \in S_t, \{\forall tt \in S_t | tt \leq t\}. \tag{3.21}$$

If more than two temperature levels are contained within a sub-volume j at time t, there exist two temperatures in the past $(\forall tt \in S_t | tt \leq t)$ having a smaller distance Δx_{tt} than the size of the sub-volume $\Delta x_{j,t}$. However, if the number of sub-volumes n_p is increased above n_p^*, inequality (3.21) holds for all sub-volumes j and time slots t and, hence, more than two temperatures within any sub-volume j are not possible. □

To calculate the pipe outflow temperature T_t^{out} in time slot t, the average of the temperatures $T_{j,t}^{out}$ of the n_p equal-sized sub-volumes j leaving the pipe in time slot t needs to be calculated like in Example 3.1. As outer approximation of the pipe outflow temperature T_t^{out} we get

$$T_t^{out} \leq \frac{1}{n_p} \sum_{j=1}^{n_p} T_{t,j}^{out} \qquad \forall t \in S_t, \tag{3.22}$$

$$T_t^{out} \geq \frac{1}{n_p} \sum_{j=1}^{n_p} T_{t,j}^{out} \qquad \forall t \in S_t. \tag{3.23}$$

As motivated in Example 3.1 and proven in Theorem 3.2, an increase of the number of sub-volumes n_p leads to an improved precision of this outer approximation. Thus, the outer approximation for variable-dependent time delays not considering thermal losses presented in (3.10) to (3.23) can serve as a dual in a primal-dual global optimization scheme as introduced in Section 2.2.

Theorem 3.2 (Increasing tightness of outer approximation)
If the number of sub-volumes approaches infinity $(n_p \to \infty)$*, the approximation of the pipe outflow temperature presented in this section approaches its real value* T_t^{out}.

Proof:
To prove Theorem 3.2, we note that the pipe inflow temperature T_t^{in} only changes at the beginning of time slots t as a discrete time modeling approach is used. Hence, at every time slot t, the number of past inflow temperatures is limited to the number of past time slots $t-1$. Introducing a variable for the number of past inflow temperatures $n_{T,t}$ in time slot t we can state accordingly

$$n_{T,t} \leq t < \infty \qquad \forall t \in S_t. \tag{3.24}$$

In reality, the number of past inflow temperatures $n_{T,t}$ is even more restricted, as there are physical limitations on mass flows and velocity of the transport medium. The number of possible temperatures inside a sub-volume leaving the pipe is bound by the same number $n_{T,t}$, as only past inflow temperatures can be inside a sub-volume leaving the pipe.

Increasing the number of sub-volumes n_p we will always reach a situation where

$$n_p > n_{T,t}. \tag{3.25}$$

Hence, there are $n_p - n_{T,t}$ more sub-volumes than possible temperatures and for $n_p - n_{T,t}$ sub-volumes there is at most one temperature possible. Thus, for these $n_p - n_{T,t}$ sub-volumes the sub-volume outflow temperature $T_{t,j}^{out}$ is known exactly. Having a limited $n_{T,t}$ as shown in (3.24) for large n_p we get

$$n_p - n_{T,t} \gg n_{T,t}. \tag{3.26}$$

Recalling (3.22) and (3.23), the overall outflow temperature T_t^{out} in time slot t is the average of the temperatures of sub-volumes $T_{t,j}^{out}$.

$$T_t^{out} = \frac{1}{n_p} \sum_{j=1}^{n_p} T_{t,j}^{out} \qquad \forall t \in S_t \tag{3.27}$$

Increasing the number of sub-volumes n_p, the at least $n_p - n_{T,t}$ sub-volumes with exactly known outflow temperatures $T_{t,j}^{out}$ get increasingly more weight in (3.27) than the at most $n_{T,t}$ sub-volumes with approximated outflow temperatures $T_{t,j}^{out}$. Thus, the outer approximation is getting more tight and exact with increasing n_p and

$$\lim_{n_p \to \infty} \frac{1}{n_p} \sum_{j=1}^{n_p} T_{t,j}^{out} = T_t^{out} \qquad \forall t \in S_t. \tag{3.28}$$

□

So far, (3.10) to (3.13) and (3.17) to (3.18) approximate the outflow temperature of one sub volume without consideration of thermal losses. To integrate thermal losses, these equations need to be combined with (2.15). Thereby, (3.10) becomes

$$T_{t,j}^{out} \leq \underbrace{T^{amb} + \left(T_{t-\tau}^{in} - T^{amb}\right) e^{-\frac{4k_V}{\rho c_p d}\Delta t \cdot \tau}}_{:=\widetilde{T}_{t-\tau}^{in}} + \left(2 - b_{\tau,j,t} - b_{\tau,j+1,t}\right) \cdot T^U$$
$$\forall j \in S_j, \forall t \in S_t, \forall \tau \in S_\tau. \tag{3.29}$$

As in (2.15), k_V, c_p, ρ and d are parameters of pipe and transport medium, Δt is the length of one time slot and τ is the discretized time delay coming with $b_{\tau,j,t}$. For the remaining (3.11) to (3.13) and (3.17) to (3.18) approximating the sub-volume outflow temperatures, heat losses can be integrated accordingly with $\widetilde{T}_{t-\tau}^{in}$ as defined in (3.29):

$$T_{t,j}^{out} \geq \widetilde{T}_{t-\tau}^{in} - \left(2 - b_{\tau,j,t} - b_{\tau,j+1,t}\right) \cdot T^U, \forall t \in S_t, \forall j \in S_j \qquad \forall \tau \in S_\tau \tag{3.30}$$

$$T_{t,j}^{out} \leq \max\left(\widetilde{T}_{t-\tau}^{in}, \widetilde{T}_{t-\tau+1}^{in}\right) + \left(2 - b_{\tau,j,t} - b_{\tau-1,j+1,t}\right) \cdot T^U$$
$$\forall t \in S_t, \forall j \in S_j, \forall \tau \in S_\tau \tag{3.31}$$

$$T_{t,j}^{out} \geq \min\left(\widetilde{T}_{t-\tau}^{in}, \widetilde{T}_{t-\tau+1}^{in}\right) - \left(2 - b_{\tau,j,t} - b_{\tau-1,j+1,t}\right) \cdot T^U$$
$$\forall t \in S_t, \forall j \in S_j, \forall \tau \in S_\tau \tag{3.32}$$

$$T_{t,j}^{out} \leq \max\left(\widetilde{T}_{t-\tau}^{in}, \widetilde{T}_{t-\tau+1}^{in}, \widetilde{T}_{t-\tau+2}^{in}\right) + \left(2 - b_{\tau,j,t} - b_{\tau-2,j+1,t}\right) \cdot T^U$$
$$\forall t \in S_t, \forall j \in S_j, \forall \tau \in S_\tau \quad (3.33)$$

$$T_{t,j}^{out} \geq \min\left(\widetilde{T}_{t-\tau}^{in}, \widetilde{T}_{t-\tau+1}^{in}, \widetilde{T}_{t-\tau+2}^{in}\right) - \left(2 - b_{\tau,j,t} - b_{\tau-2,j+1,t}\right) \cdot T^U$$
$$\forall t \in S_t, \forall j \in S_j, \forall \tau \in S_\tau. \quad (3.34)$$

The min and max functions are implemented using (3.14) to (3.16). Consideration of heat losses does not change (3.22) and (3.23), the combination of sub-volume outflow temperatures to the pipe outflow temperature T_t^{out}.

Combining (3.6) to (3.9), (3.22) and (3.23) as well as (3.29) to (3.34) we get an outer-approximation of the pipe outflow temperature as MILP problem using the basic concepts of the node method (Section 2.1.2). Thus, they can replace the equation for temperature propagation and loss (2.15) in the dual problem of a global optimization scheme for heating grids. To complete the MILP formulation of the dual problem, the bilinear terms in the node method required for temperature mixing in nodes (2.8) and change of inner energy at producers and loads (2.9) are modeled using multiparametric disaggregation introduced in Section 2.2.2 and the unit commitment equations introduced in Section 2.1.1 are added. The resulting cost minimization problem of heating grid operations is summarized in Model 3.1.

Remark: If the overall cost is minimized or profit or energy efficiency are maximized, the MILP solver will favor to produce less thermal losses. Thus, for heating grids the optimizer will maximize the pipe outflow temperature to avoid thermal losses and, hence, (3.22), (3.29), (3.31) and (3.33) are sufficient. If the approach is adapted for cooling grids only (3.23), (3.30), (3.32) and (3.34) need to be used as the optimizer will try to reduce the pipe outflow temperature to avoid thermal losses. (3.6) to (3.9), (3.22) and (3.23) are needed for both cases, heating and cooling grids.

Model 3.1: Dual problem of multiparametric delay modeling

Objective function

$$\min \sum_{t \in S_t} \left(price_t^{el,buy} \cdot p_t^{buy} - price_t^{el,sell} \cdot p_t^{sell}\right)$$
$$+ \sum_{t \in S_t} \sum_{i \in S_g} \left(C_i^{const} \cdot n_{i,t} + C_i^{el,var} \cdot p_{i,t} + C_i^{heat,var} \cdot q_{i,t}\right)$$

Generation and load constraints

$$n_{i,t} \cdot P_i^L \leq p_{i,t} \leq n_{i,t} \cdot P_i^U \qquad \forall i \in S_g, \forall t \in S_t$$
$$n_{i,t} \cdot Q_i^L \leq q_{i,t} \leq n_{i,t} \cdot Q_i^U \qquad \forall i \in S_g, \forall t \in S_t$$

$$p_t^{demand} = \sum_{i \in S_g} p_{i,t} + p_t^{buy} - p_t^{sell} \qquad \forall t \in S_t$$

$$q_{i,t} = c_p \dot{m}_{i,t} \left(T_{i,t}^{supply} - T^{return} \right) \qquad \forall i \in S_g \cup S_d, \forall t \in S_t$$

The constraints for the feasible regions of the CHPs need to be added matching their type (e.g. (2.3) for CHPs with extraction condensing turbine).

Node balances

$$\sum_{i \in inflow} \dot{m}_{i,t} = \sum_{i \in outflow} \dot{m}_{i,t} \qquad \forall t \in S_t$$

$$T_t^{out} = \frac{\sum_{i \in inflow} \left(T_{i,t}^{in} \cdot \dot{m}_{i,t} \right)}{\sum_{i \in inflow} \dot{m}_{i,t}} \qquad \forall t \in S_t$$

Assignment of binary delay

$$\Delta x_t = \frac{\Delta t}{\rho \pi \left(\frac{d}{2} \right)^2} \dot{m}_t \qquad \forall t \in S_t$$

$$\sum_{\tau=0}^{t} b_{\tau,j,t} = 1 \qquad \forall t \in S_t, \forall j \in S_j$$

$$L - \frac{j-1}{n_p} \Delta x_t \leq \sum_{tt=t-\tau}^{t-1} \Delta x_{tt} + (1 - b_{\tau,j,t}) \cdot M^{\star} \qquad \forall t \in S_t, \forall j \in S_j, \forall \tau \in S_\tau$$

$$L - \frac{j-1}{n_p} \Delta x_t \geq \sum_{tt=t-\tau+1}^{t-1} \Delta x_{tt} - (1 - b_{\tau,j,t}) \cdot M^{\star} \qquad \forall t \in S_t, \forall j \in S_j, \forall \tau \in S_\tau$$

Pipe outflow temperature calculation

$$T_{t,j}^{out} \leq \underbrace{T^{amb} + \left(T_{t-\tau}^{in} - T^{amb} \right) e^{-\frac{4k_V}{\rho c_p d} \Delta t \cdot \tau}}_{:= \widetilde{T}_{t-\tau}^{in}} + (2 - b_{\tau,j,t} - b_{\tau,j+1,t}) \cdot T^U$$

$$\forall j \in S_j, \forall t \in S_t, \forall \tau \in S_\tau$$

$$T_{t,j}^{out} \leq \max \left(\widetilde{T}_{t-\tau}^{in}, \widetilde{T}_{t-\tau+1}^{in} \right) + (2 - b_{\tau,j,t} - b_{\tau-1,j+1,t}) \cdot T^U$$

$$\forall t \in S_t, \forall j \in S_j, \forall \tau \in S_\tau$$

$$T_{t,j}^{out} \leq \max\left(\widetilde{T}_{t-\tau}^{in}, \widetilde{T}_{t-\tau+1}^{in}, \widetilde{T}_{t-\tau+2}^{in}\right) + \left(2 - b_{\tau,j,t} - b_{\tau-2,j+1,t}\right) \cdot T^{U}$$
$$\forall t \in S_t, \forall j \in S_j, \forall \tau \in S_\tau$$

$$T_t^{out} \leq \frac{1}{n_p} \sum_{j=1}^{n_p} T_{t,j}^{out} \qquad \forall t \in S_t$$

All multiplications and divisions of variables are modeled using multiparametric disaggregation (Section 2.2.2).

Function $c \leq \max(a, b)$ is enforced using the following equations.

$$a - b = \delta^+ - \delta^-$$
$$c \leq b + \delta^+$$

3.2 Primal problems

In addition to the dual problem presented in Section 3.1 a model for the primal problem is required to complete a primal-dual global optimization scheme. Because of the variable-dependent time delay, it is unfortunately not possible to directly implement the node method as presented in Section 2.1.2 for non-linear optimization solvers like IPOPT [WB06]. Thus, to get a fast and easy to solve NLP as primal problem as in [TCM12], in the following two different approaches are introduced to model the variable-dependent delay. Both are first published in [MLH19] and adapt prior work [vFR$^+$17, BHW17]. The approach presented in Section 3.2.1 uses first order differential equations to model the transport delay. Section 3.2.2 presents a refinement of this approach combining a static time delay calculated using the node method with a variable delay model based on first-order differential equations. This leads to a second formulation for the primal problem. The bilinear terms remain for the primal problems presented in the following as they can be solved with local NLP solvers.

3.2.1 Finite-volumes pipe formulation

One common way to approximate dynamic time delays is to combine a series of first-order differential equations (PT1 elements) [vFR$^+$17]. This approximation is getting more exact if the number of first-order differential equations increases. Applying this approach to the mass-flow dependent transport delay of a pipe, each first-order differential equation represents the energy balance of a volume element in the pipe.

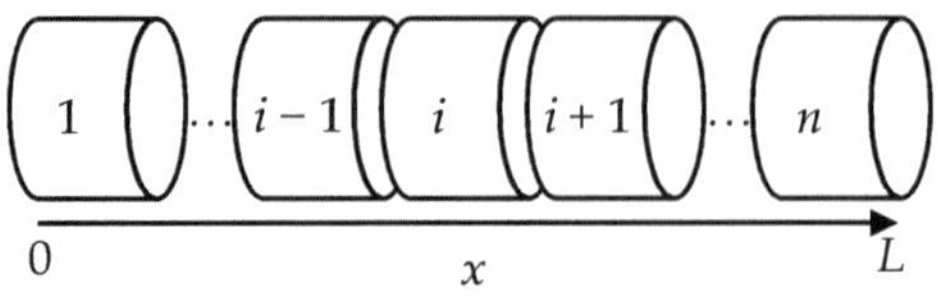

Figure 3.3: Scheme of a pipe of length L split into n finite volumes i.

As shown in Figure 3.3 the pipe is split into a finite number of n volume elements i of equal size. Assuming perfect mixing, the water inside each of these volume elements i is at only one average temperature $T^{av}_{i,t}$. Hence, recalling (2.13) the thermal loss $q^{loss}_{i,t}$ of one volume element i in a time slot t becomes

$$q^{loss}_{i,t} = \frac{k_V \pi d L}{n} \left(T^{av}_{i,t} - T^{amb} \right) \qquad \forall i \in S_i, \forall t \in S_t. \tag{3.35}$$

As for (2.13), k_V, d and L are parameters of the pipe and T^{amb} is the ambient temperature. The set S_i denotes all volume elements. The transport of thermal energy $\dot{Q}^{convection}_{i,t}$ from one volume element $i-1$ to the next volume element i is based on convection at mass flow $\dot{m}_t$ with the transport medium having a specific heat capacity c_p and, thus, can be described with

$$\dot{Q}^{convection}_{i,t} = c_p \dot{m}_t \left(T^{av}_{i-1,t} - T^{av}_{i,t} \right) \qquad \forall i \in S_i, \forall t \in S_t. \tag{3.36}$$

The thermal energy balance for one volume element i is

$$\dot{Q}_{i,t} = \dot{Q}^{convection}_{i,t} - q^{loss}_{i,t} \qquad \forall i \in S_i, \forall t \in S_t. \tag{3.37}$$

Combining the above equations with (2.9) for the inner energy change $\dot{Q}_{i,t}$ we get the energy balance of one volume element i with mass $m = n/(\rho\pi(d/2)^2 L)$:

$$c_p m \cdot \left(T^{av}_{i,t+1} - T^{av}_{i,t} \right) = c_p \dot{m}_t \left(T^{av}_{i-1,t} - T^{av}_{i,t} \right) - \frac{k_V \pi d L}{n} \left(T^{av}_{i,t} - T^{amb} \right) \qquad \forall i \in S_i, \forall t \in S_{t<n_t}. \tag{3.38}$$

With some minor reformulations, the resulting first-order difference equation is

$$T^{av}_{i,t+1} = T^{av}_{i,t} + \frac{n}{c_p \rho \pi \left(\frac{d}{2}\right)^2 L} \left(c_p \dot{m}_t \left(T^{av}_{i-1,t} - T^{av}_{i,t} \right) - \frac{k_V \pi d L}{n} \left(T^{av}_{i,t} - T^{amb} \right) \right) \qquad \forall i \in S_i, \forall t \in S_{t<n_t}. \tag{3.39}$$

For a flow direction from first volume element $i = 1$ to last volume element $i = n$ the outflow temperature of the pipe T^{out}_t is the temperature $T_{n,t}$ of the last volume element n. The inflow temperature of the pipe T^{in}_t replaces the temperature of the previous volume element $T^{av}_{i-1,t}$ in the energy balance equation (3.39) for the first volume element $i = 1$.

The resulting model formulation is shown in Model 3.2. It allows to find fast and reliable solutions using local solvers for NLPs like IPOPT [WB06]. But as discussed in [Trö99] introducing a finite number of volume elements, each at only one temperature level, leads to increased averaging of the temperatures and, thus, smoothing of the delayed variable. Thus, it is less exact than the node method. This effect decreases with increasing number of volume elements n [Trö99]. However, a higher number of volume elements n increases problem complexity and solution time.

Model 3.2: Primal problem using finite-volumes pipe model

Objective function

$$\min \sum_{t\in S_t} \left(price_t^{el,buy} \cdot p_t^{buy} - price_t^{el,sell} \cdot p_t^{sell} \right) + \sum_{t\in S_t} \sum_{i\in S_g} \left(C_i^{const} \cdot n_{i,t} + C_i^{el,var} \cdot p_{i,t} + C_i^{heat,var} \cdot q_{i,t} \right)$$

Generation and load constraints

$$n_{i,t} \cdot P_i^L \leq p_{i,t} \leq n_{i,t} \cdot P_i^U \qquad \forall i \in S_g, \forall t \in S_t$$

$$n_{i,t} \cdot Q_i^L \leq q_{i,t} \leq n_{i,t} \cdot Q_i^U \qquad \forall i \in S_g, \forall t \in S_t$$

$$p_t^{demand} = \sum_{i\in S_g} p_{i,t} + p_t^{buy} - p_t^{sell} \qquad \forall t \in S_t$$

$$q_{i,t} = c_p \dot{m}_{i,t} \left(T_{i,t}^{supply} - T^{return} \right) \qquad \forall i \in S_g \cup S_d, \forall t \in S_t$$

The constraints for the feasible regions of the CHPs need to be added matching their type (e.g. (2.3) for CHPs with extraction condensing turbine).

Node balances

$$\sum_{i\in inflow} \dot{m}_{i,t} = \sum_{i\in outflow} \dot{m}_{i,t} \qquad \forall t \in S_t$$

$$T_t^{out} = \frac{\sum_{i\in inflow} \left(T_{i,t}^{in} \cdot \dot{m}_{i,t} \right)}{\sum_{i\in inflow} \dot{m}_{i,t}} \qquad \forall t \in S_t$$

Temperature propagation in pipes

$$T_{1,t+1}^{av} = T_{1,t}^{av} + \frac{n}{c_p \rho \pi \left(\frac{d}{2}\right)^2 L} \left(c_p \dot{m}_t \left(T_t^{in} - T_{1,t}^{av} \right) - \frac{k_V \pi d L}{n} \left(T_{1,t}^{av} - T^{amb} \right) \right) \qquad \forall t \in S_{t<n_t}$$

$$T_{i,t+1}^{av} = T_{i,t}^{av} + \frac{n}{c_p \rho \pi \left(\frac{d}{2}\right)^2 L} \left(c_p \dot{m}_t \left(T_{i-1,t}^{av} - T_{i,t}^{av}\right) - \frac{k_V \pi d L}{n} \left(T_{i,t}^{av} - T^{amb}\right)\right)$$

$$\forall i \in S_{i>1}, \forall t \in S_{t<n_t}$$

$$T_t^{out} = T_{n,t} \qquad \forall t \in S_t$$

3.2.2 Hybrid pipe formulation

To get a more accurate primal problem with still acceptable computational performance a combination of a pipe model with a static transport delay, calculated using the node method (Section 2.1.2), and a flexible transport delay, modeled using first-order differential equations (as the model presented in Section 3.2.1), is introduced in the following. This concept is based on [BHW17]. Because in real-world applications there is always an upper bound on mass flow and velocity, there is a minimum of the transport delay. This static minimum transport delay τ^{min} can be calculated based on pipe length L and upper bound of velocity v^U or mass flow $\dot{m}^U$ using

$$\tau^{min} = \frac{L}{v^U} = L \frac{\rho \pi \left(\frac{d}{2}\right)^2}{\dot{m}^U}. \tag{3.40}$$

If the upper bound of mass flow $\dot{m}^U$ is used for calculation, pipe diameter d and density of the transport medium ρ are required as additional parameters.

The overall transport delay τ_t in a time slot t is split into this static minimum part τ^{min} and a flexible delay τ_t^{flex} which is modeled using first-order differential equations. Hence, we arrive at a hybrid pipe model as shown in Figure 3.4 and the overall time delay τ_t is calculated with

$$\tau_t = \tau^{min} + \tau_t^{flex}. \tag{3.41}$$

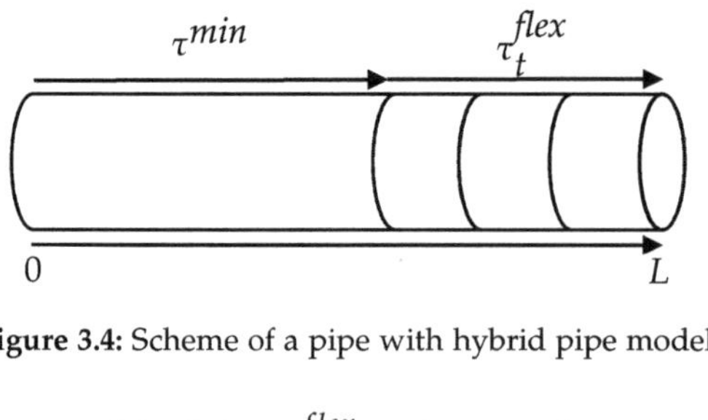

Figure 3.4: Scheme of a pipe with hybrid pipe model

Modeling the remaining variable delay τ_t^{flex} with a similar scheme as shown in Section 3.2.1 the maximum mass flow $\dot{m}^U$ needs to be subtracted from the mass flow $\dot{m}_t$ in the first-order differential equation in (3.39) resulting in

$$\frac{\rho\pi\left(\frac{d}{2}\right)^2 L}{n}\left(\frac{1}{\dot{m}_t}-\frac{1}{\dot{m}^U}\right)\left(T^{av}_{i,t+1}-T^{av}_{i,t}\right)=\left(T^{av}_{i-1,t}-T^{av}_{i,t}\right)$$
$$-\frac{k_V\pi d L}{c_p n}\left(\frac{1}{\dot{m}_t}-\frac{1}{\dot{m}^U}\right)\left(T^{av}_{i,t}-T^{amb}\right) \qquad \forall i\in S_i,\forall t\in S_{t<n_t}. \tag{3.42}$$

Compared to the finite-volumes model presented in Section 3.2.1, the inflow temperature $T^{av}_{i-1,t}$ for the first volume element $i=1$ is here replaced with the pipe inflow temperature T^{in}_t delayed by the constant minimum transport delay τ^{min}. If heat losses are considered we arrive at a formulation as in (2.15) and get

$$T^{av}_{i-1,t}=T^{amb}+\left(T^{in}_{t-\tau^{min}}-T^{amb}\right)e^{-\frac{4\,k_V}{c_p\rho\cdot d}\tau^{min}} \qquad \forall t\in S_t, i=1. \tag{3.43}$$

The average temperature $T_{n,t}$ of the last volume element n remains the outflow temperature T^{out}_t of the pipe in time slot t. The resulting model formulation for a primal problem using the hybrid pipe model is presented in Model 3.3.

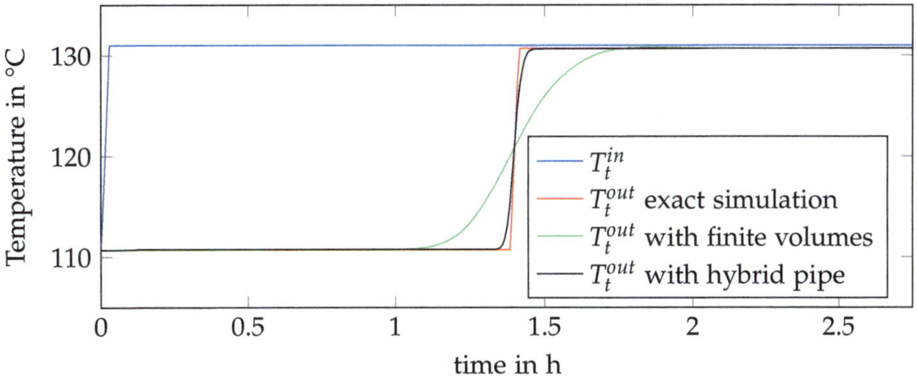

Figure 3.5: Simulation results comparing finite-volumes and hybrid pipe models with an exact modeling of the transport delay for a pipe with 13.5 km length and 0.7 m diameter. 100 volume elements are used for the finite volumes model. 50 volume elements and a maximum flow of 3 m/s are used with the hybrid pipe model.

This hybrid pipe model enables a more accurate representation of the pipeline dynamics than the finite-volumes model presented in Section 3.2.1, as the smoothing of the delay only acts on τ^{flex}_t. The temperature propagation including losses can be exactly calculated for the static minimum delay τ^{min}. This advantage of the hybrid pipe model is illustration in a simulation in Figure 3.5. However, depending on the combination of discretization time, minimum delay time and pipe length the advantage of the hybrid pipe model to the finite-volumes model (Section 3.2.1) varies.

Model 3.3: Primal problem using the hybrid pipe model formulation

Objective function

$$\min \sum_{t\in S_t} \left(price_t^{el,buy} \cdot p_t^{buy} - price_t^{el,sell} \cdot p_t^{sell} \right) + \sum_{t\in S_t} \sum_{i\in S_g} \left(C_i^{const} \cdot n_{i,t} + C_i^{el,var} \cdot p_{i,t} + C_i^{heat,var} \cdot q_{i,t} \right)$$

Generation and load constraints

$$n_{i,t} \cdot P_i^L \leq p_{i,t} \leq n_{i,t} \cdot P_i^U \qquad \forall i \in S_g, \forall t \in S_t$$

$$n_{i,t} \cdot Q_i^L \leq q_{i,t} \leq n_{i,t} \cdot Q_i^U \qquad \forall i \in S_g, \forall t \in S_t$$

$$p_t^{demand} = \sum_{i\in S_g} p_{i,t} + p_t^{buy} - p_t^{sell} \qquad \forall t \in S_t$$

$$q_{i,t} = c_p \dot{m}_{i,t} \left(T_{i,t}^{supply} - T^{return} \right) \qquad \forall i \in S_g \cup S_d, \forall t \in S_t$$

The constraints for the feasible regions of the CHPs need to be added matching their type (e.g. (2.3) for CHPs with extraction condensing turbine).

Node balances

$$\sum_{i\in inflow} \dot{m}_{i,t} = \sum_{i\in outflow} \dot{m}_{i,t} \qquad \forall t \in S_t$$

$$T_t^{out} = \frac{\sum_{i\in inflow} \left(T_{i,t}^{in} \cdot \dot{m}_{i,t} \right)}{\sum_{i\in inflow} \dot{m}_{i,t}} \qquad \forall t \in S_t$$

Temperature propagation in pipes

$$\tau^{min} = L \frac{\rho \pi \left(\frac{d}{2}\right)^2}{\dot{m}^U}$$

$$T_{i-1,t}^{av} = T^{amb} + \left(T_{t-\tau^{min}}^{in} - T^{amb} \right) e^{-\frac{4\,k_V}{c_p \rho \cdot d} \tau^{min}} \qquad \forall t \in S_t, i = 1$$

$$\frac{\rho \pi \left(\frac{d}{2}\right)^2 L}{n} \left(\frac{1}{\dot{m}_t} - \frac{1}{\dot{m}^U} \right) \left(T_{i,t+1}^{av} - T_{i,t}^{av} \right) = \left(T_{i-1,t}^{av} - T_{i,t}^{av} \right) - \frac{k_V \pi d L}{c_p n} \left(\frac{1}{\dot{m}_t} - \frac{1}{\dot{m}^U} \right) \left(T_{i,t}^{av} - T^{amb} \right) \qquad \forall i \in S_i, \forall t \in S_{t<n_t}$$

$$T_t^{out} = T_{n,t} \qquad \forall t \in S_t$$

3.3 Global optimization algorithm

Given the dual problem presented in Section 3.1 and one of the primal problems presented in Section 3.2 we can now combine them to a primal-dual global optimization algorithm for problems with variable-dependent time delays. For heating grids, there are non-convex bilinear terms besides the variable-dependent time delay. Therefore, to get a global optimization scheme for heating grids we extend the global optimization scheme of multiparametric disaggregation (see Section 2.2.2) with the dual and primal models presented in this chapter.

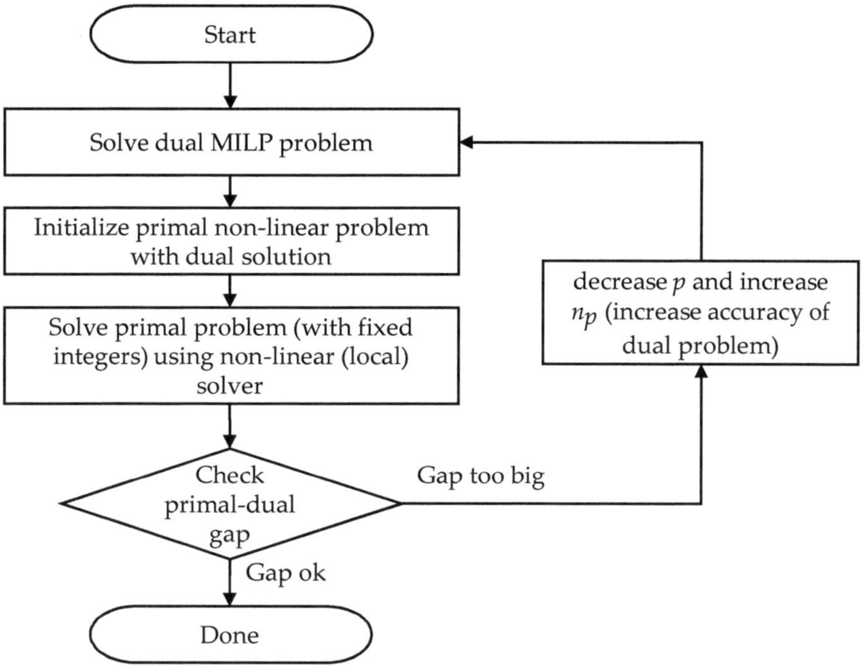

Figure 3.6: Iterative solution approach of multiparametric delay modeling

This extension of the iterative scheme of multiparametric disaggregation is shown in Figure 3.6. The dual problem combines the outer approximations of multiparametric disaggregation (Section 2.2.2) and multiparametric delay modeling (Section 3.1). Thus, an increase of the model precision requires a higher precision of approximation of the bilinear terms (by a decrease of p) and a higher precision of approximation of the variable-dependent delays (by an increase of n_p). One of the formulations presented in Section 3.2 is used as primal problem modeling the variable-dependent time delay.

The number of iterations needed to achieve an acceptable gap between the solutions of the primal and the dual problem depends on heating grid size and complexity. In theory, with unlimited computational power and time the solution gap approaches zero. In reality, however, a gap will remain in most cases. For complex problems, a gap of about 1 % is usually accepted being one order of magnitude above the default gap of off-the-shelf MILP solvers (usually 0.1 %) [Gur16].

3.4 Summary of chapter

This chapter presented "multiparametric delay modeling", the first approach proving global optimality of its solution for optimization of district heating grids. "Multiparametric delay modeling" is a deterministic global optimization scheme based on the decomposition of the original problem into a primal and a dual problem that are easier to solve than the original problem. First, Section 3.1 introduced a novel dual model formulation to represent variable-dependent delays. It approximates the bounds of the pipe outflow temperature by introducing binary variables representing the transport time in every time slot. Thus, the resulting problem is a MILP problem which can be solved with a variety of efficient off-the-shelf solvers like [Gur16]. The outflowing volume per time slot is split into sub-volumes to refine this approximation. Theorem 3.1 and Theorem 3.2 prove that this outer approximation becomes more and more accurate with increasing number of sub-volumes.

In Section 3.2, two different primal models were introduced which can be solved with off-the-shelf NLP solvers using e.g. an interior point method [WB06]. In the first primal model, the pipe is split into a number of equal-sized finite volumes. For each volume a constant temperature and perfect mixing are assumed allowing to represent the thermal dynamics of the volume in one energy balance. The second primal model combines this finite volumes model with a static time delay leading to a hybrid pipe model. This hybrid pipe model allows a more accurate representation of the transport delay than the pure finite volumes approach of the first primal model.

Finally, Section 3.3 established an extension of the global optimization algorithm of multiparametric disaggregation (Section 2.2.2), which enables to globally optimize the fully dynamic problem of heating grid optimization. For this, multiparametric disaggregation for global optimization of bilinear terms is combined with "multiparametric delay modeling" for global optimization of variable-dependent time delays. The dual problem presented in Model 3.1 and one of the primal problems (Model 3.2 or Model 3.3) are used in this global optimization scheme.

4 Optimization with hybrid discrete-continuous time grid

As the solution algorithm in the previous chapter is an iterative optimization scheme with multiple solver runs, solution times can be quite long. This chapter aims at finding an exact model which enables optimization in one solver run, promising faster solution times than an iterative solution approach. Recent advances in modeling in other areas help to achieve this goal. Here, we extend formulations from pipeline scheduling for refined, liquid petroleum products (as presented in Section 2.3) to become suitable for heating grid optimization. This approach was developed in a collaboration with Pedro Castro from University of Lisbon, Portugal and is published in [MC20]. It adapts the pipeline scheduling model presented in [Cas17].

As presented in Section 2.3, there are many similarities between pipeline scheduling for refinery products and optimization of heating grids considering the grid dynamics. Limiting the choice of supply temperature to a discrete set of temperature levels allows to use pipeline scheduling approaches for district heating grid optimization. In addition, a hybrid discrete-continuous time grid is introduced adapting the continuous time formulation of pipeline scheduling models to time discrete energy prices.

4.1 Hybrid discrete-continuous time grid

The choice of an appropriate time representation is crucial for scheduling models [HMB+14, MHI+15]. Looking at the literature review in Section 2.1.2, Section 2.1.4 and Section 2.1.5, discrete time models are dominating past publications on heating grid optimization. However, for pipeline scheduling continuous time representations have shown to be very efficient to model flow dynamics (see Section 2.3).

A combination of the two is presented in [WL18] where a hybrid discrete-continuous time grid models thermal dynamics of a district heating grid. In particular, fixed time points are used at every border between periods of constant demand and prices and combined with floating time points to account for the variability of the time delay. As the number of floating points between two fixed time points depends on the thermal grid setup, an upper bound of two per thermal storage tank in the system is proven in [WL18], but no model formulation suitable for optimization is presented. Therefore, the hybrid-discrete continuous time model for district heating grid optimization is introduced in the following as first presented in [MC20]. In this model thermal storage tanks are not considered, but the dynamics of the district heating grid are used as thermal storage. Heuristically, one floating point is defined between two fixed time points based on optimal performance in a small case study. In systems with a larger number of pipes and nodes, a higher number of floating time points between two fixed time points might be needed, increasing the complexity of the formulation.

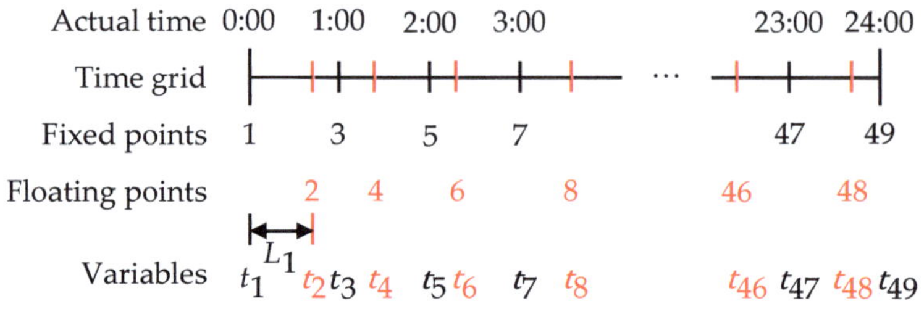

Figure 4.1: Hybrid discrete-continuous time grid

Figure 4.1 shows the hybrid discrete-continuous time grid. All odd-numbered time points are located at hour marks tfx_t and are thus fixed time points $t \in S_t^{fx}$. All even-numbered time points are floating time points allowed to move freely between the neighboring fixed time points. For one day or 24 hours we get a total of $n_t = 49$ time points (see Figure 4.1) and accordingly 48 time slots as every hour is split into two time slots by the floating time point. A big advantage of this time representation in comparison to a continuous time representation is that the electricity price $price_t^{el}$ for each time slot is known a priori and a complicated assignment of variables to time slots as for example in [HH13] is not needed. The same applies to the heat demand q_t^{demand} which is forecasted with hourly changing values.

Let us introduce non-negative real variables to model this hybrid discrete-continuous time grid. t_t represents the actual time at time point t, whereas L_t represents the duration of the time slot.

$$t_{t+1} = t_t + L_t \qquad \forall t \in S_{t<n_t} \tag{4.1}$$

allows to link the two. All even time points are fixed time points tfx_t which is enforced by

$$t_t = tfx_t \qquad \forall t \in S_t^{fx}. \tag{4.2}$$

All odd-numbered time points are floating time points $t \notin S_t^{fx}$ which are allowed to vary between the neighboring time points in interval $[t_{t-1}, t_{t+1}]$. This can be guaranteed by limiting the time slot length L_t to one hour, the distance between two fixed time points. As a time point is needed, whenever there is a new batch of water entering or leaving a pipe and this does usually not happen at the hour marks, the floating time points allow an exact representation of varying in- and outflows. In some cases there will be no need for a floating time point between the hour marks e.g. when there is no change in the in- or outflow of the pipe. This leads to dummy time slots with $L_{t-1} = 0$ or $L_t = 0$ as the optimization chooses t_t equal to either t_{t-1} or t_{t+1}.

Remark: As explained in [MC20], choosing a continuous time grid where all hour marks need to be assigned to a time point and all time points are floating time points as proposed in [CHG09] does not lead to superior solutions. At least $n_t = 26$ time points are needed to get a feasible problem, being one above the minimum number for one time slot per hour. But in comparison to the formulation using the hybrid discrete-continuous time grid, this continuous time formulation has a larger problem size, worse results and orders of magnitude longer solution times.

4.2 Supply node

Besides the hybrid discrete-continuous time grid another novelty in [MC20] is the discretization of temperature levels that enables to adapt the product-centric pipeline scheduling formulation [Cas17] to district heating grid optimization. The following formulations are introduced for one CHP, but adding more CHPs is straightforward by adding an additional index to all CHP variables and parameters.

In comparison to the basic unit commitment model presented in Section 2.1.1 the binary variable $n_{i,t}$ denoting the running state of CHP i is extended with one additional index l, being the temperature level currently produced by the CHP. Thus, for each CHP we now have a binary variable $X^C_{l,t} = 1$ if in time slot t the CHP is producing water at temperature level l. As at most one temperature can be produced in a time slot, we get

$$\sum_{l \in S_l} X^C_{l,t} \leq 1 \qquad \forall t \in S_{t<n_t}. \tag{4.3}$$

The set S_l contains all possible temperature levels l. When the CHP is producing water at temperature level l in time slot t, the volume $F^C_{l,t}$ (in m^3) enters the pipe, being a non-negative real variable with lower and upper bounds f^L and f^U. In comparison to the lower and upper bounds Q^L and Q^U on thermal energy generation of the CHP presented in Section 2.1.1, these are bounds on the volume injected by the CHP which can be calculated for example based on minimum and maximum mass flow or velocity:

$$f^L X^C_{l,t} \leq F^C_{l,t} \leq f^U X^C_{l,t} \qquad \forall l \in S_l, \forall t \in S_{t<n_t}. \tag{4.4}$$

The change of inner energy of the transport medium at the producer (see (2.9)) needs to account for the different temperature levels l as well. The thermal energy volume q_t produced in time slot t is calculated based on supply temperature levels T^{supply}_l and return temperature T^{return} (assumed to be constant) with

$$q_t = \sum_l F^C_{l,t} \cdot \rho \cdot c_p \cdot \left(T^{supply}_l - T^{return}\right) \qquad \forall t \in S_{t<n_t}. \tag{4.5}$$

As before, c_p is the specific heat coefficient of the transport medium (water) and ρ is its density.

The equations for the electric generation as well as the cost function remain the same as for the basic unit commitment model (Section 2.1.1), when both time slots within one hour are assigned to the parameters of this hour. The equations linking electric and thermal output of the CHP, however, need to consider the variable length of the time slots. Thus, formulation (2.3) for the feasible region of CHP operation with extraction condensing turbine is not suitable. To match the real feasible operation region (Figure 2.1) the inequalities (2.3) are adapted to

$$\begin{aligned} \frac{p_t}{P^U} &\leq \alpha_1 \cdot L_t - \beta_1 \frac{q_t}{Q^U} && \forall t \in S_t, \\ \frac{p_t}{P^U} &\geq \alpha_2 \cdot L_t - \beta_2 \frac{q_t}{Q^U} && \forall t \in S_t, \\ \frac{p_t}{P^U} &\geq -\alpha_3 \cdot L_t + \beta_3 \frac{q_t}{Q^U} && \forall t \in S_t \end{aligned} \tag{4.6}$$

with parameters α_1, α_2, α_3, β_1, β_2 and β_3 adjusting slope and offset of the region boundaries.

4.3 Demand node

Similar to the supply node, let us introduce a binary variable $X_{l,t}^D$ being 1, if the water arriving at the demand node is at supply temperature level l in time slot t. Due to the transport delay in the pipe, this is not necessarily the supply temperature level at the CHP in time slot t. At most one temperature level l can arrive at the demand node in one time slot t:

$$\sum_l X_{l,t}^D \leq 1 \qquad \forall t \in S_{t<n_t}. \tag{4.7}$$

The volume of water at supply temperature level l consumed by the demand node $F_{l,t}^D$ (in m^3) in time slot t is bound due to limitations in mass flow and velocity. Similar to the volume produced by the CHP we get

$$f^L X_{l,t}^D \leq F_{l,t}^D \leq f^U X_{l,t}^D \qquad \forall l \in S_l, \forall t \in S_{t<n_t}. \tag{4.8}$$

As the mass flow to the demand node is limited to interval $[\dot{m}^L, \dot{m}^U]$, the operating volume flow needs to stay always within these bounds as well. Without an explicit variable for the mass flow, we can ensure these bounds with

$$\sum_l \frac{F_{l,t}^D}{\dot{m}^U/\rho} \leq L_t \leq \sum_l \frac{F_{l,t}^D}{\dot{m}^L/\rho} + 1 - \sum_l X_{l,t}^D \qquad \forall t \in S_{t<n_t}. \tag{4.9}$$

This is possible, because mass and volume flow are connected via the density ρ of the transport medium. For this equation it is important that no time slot is longer than 1 hour.

The energy balance for the demand node cooling water from supply temperature level T_l^{supply} to return temperature T^{return} is explained by the change of inner energy of the transport medium based on (2.9). As for the CHP, all different possible temperature levels l need to be considered. In addition, the thermal demand needs to be scaled according to time slot length L_t to guarantee a continuous delivery of thermal energy:

$$q_t^{demand} \cdot L_t = \sum_l F_{l,t}^D \cdot \rho \cdot c_p \cdot \left(T_l^{supply} - T^{return}\right) \qquad \forall t \in S_{t<n_t}. \tag{4.10}$$

If thermal losses are to be considered, the supply temperature level T_l^{supply} in the previous equation needs to be adjusted to represent the temperature drop along the pipe. As the supply temperature is the major influence on thermal losses (2.15) each supply temperature level at the demand node must be corrected individually by introducing new parameters representing the reduced supply temperatures per temperature level at the demand node.

4.4 Pipe representation

To model the flow dynamics in a pipe, we introduce the non-negative real variable $V_{l,t}$ which represents the volume of water at temperature level l inside the pipe in time slot t. It is assumed that there is only one volume of temperature level l inside the pipe in every time slot t. In Figure 4.2 all scenarios changing this volume are shown: If the CHP is feeding a volume $F_{l,t}^C$ at temperature level l into the pipe in time slot t, the volume $V_{l,t}$ at temperature level l inside the pipe increases as shown in Figure 4.2 a). If water at temperature level l arrives at the demand node, the volume $V_{l,t}$ at temperature level l inside the pipe decreases by $F_{l,t}^D$ as shown in Figure 4.2 c). Figure 4.2 b) shows a situation, where the volume $V_{l,t}$ at temperature level l inside the pipe does not change in time slot t. This is possible as well if $V_{l,t} = 0$ or $V_{l,t} = v$ in a time slot. The respective volume balance for the volume of water at a certain temperature level l inside the pipe in time slot t is thus

$$V_{l,t} = v_l^0|_{t=1} + \left(V_{l,t-1} + F_{l,t-1}^C - F_{l,t-1}^D\right)|_{t>1} \qquad \forall l \in S_l, \forall t \in S_t. \tag{4.11}$$

Here $v_l^0|_{t=1}$ is the initial volume of water at temperature level l inside the pipe.

As water is non-compressible, the sum of all volumes inside a pipe needs to equal the overall pipe volume v in all time slots t:

$$\sum_l V_{l,t} = v \qquad \forall t \in S_t. \tag{4.12}$$

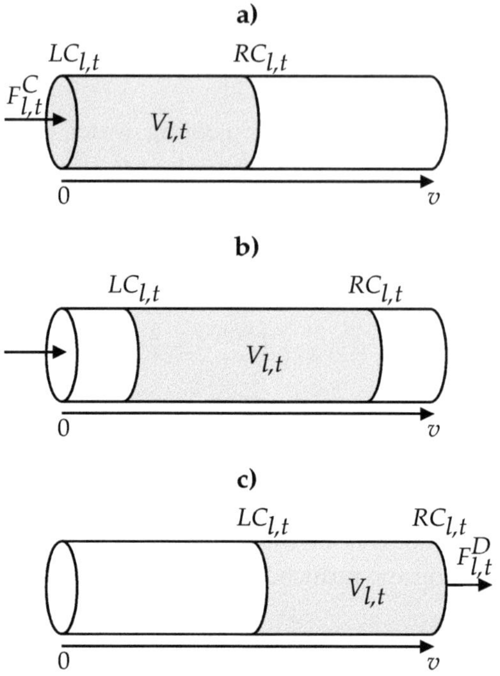

Figure 4.2: Volume $V_{l,t}$ at temperature level l a) flowing in, b) being in and c) flowing out of a pipe with overall volume v. Note that coordinates $LC_{l,t}$ and $RC_{l,t}$ are volume coordinates, not distance coordinates.

As shown in Figure 4.2, the non-negative real variables $LC_{l,t}$ and $RC_{l,t}$ describe the left and right coordinates of volume $V_{l,t}$ having temperature level l in time slot t. Note that these coordinates are not distance coordinates in m but volume coordinates in m^3 describing the position of water at temperature level l inside the pipe with respect to the overall pipe volume v. Accordingly, they are limited by pipe volume v:

$$LC_{l,t} \leq v \qquad \forall l \in S_l, \forall t \in S_t, \tag{4.13}$$

$$RC_{l,t} \leq v \qquad \forall l \in S_l, \forall t \in S_t. \tag{4.14}$$

The coordinates $LC_{l,t}$ and $RC_{l,t}$ need to be directly left and right of the volume at temperature level l and, thus, have a difference of

$$RC_{l,t} - LC_{l,t} = V_{l,t} \qquad \forall l \in S_l, \forall t \in S_t. \tag{4.15}$$

To complete the model some additional variables need to be introduced. Variables $LC^{end}_{l,t-1}$ and $RC^{end}_{l,t-1}$ describe the coordinates of water at temperature level l at the end of the last time slot $t-1$. These coordinates are different from the coordinates at the beginning of the time slot if water at a different temperature level entered or left the pipe. To link the coordinates at the end of a time slot $LC^{end}_{l,t}$ and $RC^{end}_{l,t}$ with coordinates at the beginning of the time slot $LC_{l,t}$ and $RC_{l,t}$, we use

$$LC^{end}_{l,t} = LC_{l,t} + \sum_{l' \neq l} F^{C}_{l',t} \qquad \forall l \in S_l, \forall t \in S_{t<n_t}, \tag{4.16}$$

$$RC^{end}_{l,t} = RC_{l,t} + \sum_{l' \neq l} F^{D}_{l',t} \qquad \forall l \in S_l, \forall t \in S_{t<n_t}. \tag{4.17}$$

We propagate coordinates $LC^{end}_{l,t-1}$ and $RC^{end}_{l,t-1}$ at the end of the previous time slot $t-1$ to the coordinates $LC_{l,t}$ and $RC_{l,t}$ at the beginning of time slot t using

$$LC_{l,t} = lc^0_l|_{t=1} + \left(LC^{end}_{l,t-1} - ZC_{l,t}\right)|_{t>1} \qquad \forall l \in S_l, \forall t \in S_t, \tag{4.18}$$

$$RC_{l,t} = rc^0_l|_{t=1} + \left(RC^{end}_{l,t-1} - ZC_{l,t}\right)|_{t>1} \qquad \forall l \in S_l, \forall t \in S_t. \tag{4.19}$$

In addition, these equations initialize the coordinates at the beginning of the optimization horizon to the initial values of the coordinates lc^0_l and rc^0_l for every temperature level l.

Variable $ZC_{l,t}$ is needed to reset the coordinates if water at temperature level l is not inside the pipe, such that it can enter the pipe. Hence, it is linked with the binary variable $X^P_{l,t}$ indicating if a temperature level l is currently inside the pipe:

$$ZC_{l,t} \leq f^U \cdot \left(1 - X^P_{l,t}\right) \qquad \forall l \in S_l, \forall t \in S_{t>1}. \tag{4.20}$$

Accordingly, the volume $V_{l,t}$ of water at temperature level l inside the pipe in time slot t as well as the corresponding left and right coordinates $LC_{l,t}$ and $RC_{l,t}$ need to be zero, if there is no water at temperature level l inside the pipe ($X^P_{l,t} = 0$).

$$V_{l,t} \leq v \cdot X^P_{l,t} \qquad \forall l \in S_l, \forall t \in S_{t>1} \tag{4.21}$$

$$LC_{l,t} \leq v \cdot X^P_{l,t} \qquad \forall l \in S_l, \forall t \in S_{t>1} \tag{4.22}$$

$$RC_{l,t} \leq v \cdot X^P_{l,t} \qquad \forall l \in S_l, \forall t \in S_{t>1} \tag{4.23}$$

Finally, the conditions describing when water at a temperature level l can enter or leave the pipe need to be defined. As it is assumed that only one volume of a temperature level is possible inside the pipe, the left coordinate $LC_{l,t}$ for this temperature level l has to equal zero if water at temperature level l should enter the pipe ($X^C_{l,t} = 1$) and we get

$$LC_{l,t} \leq v \cdot \left(1 - X^C_{l,t}\right) \qquad \forall l \in S_l, \forall t \in S_{t<n_t}. \tag{4.24}$$

This equation covers both cases: if there is no water at temperature level l inside the pipe ($X^P_{l,t} = 0$ and $V_{l,t} = 0$) as well as if water at temperature level l is inside the pipe ($X^P_{l,t} = 1$ and $0 \leq V_{l,t} \leq v$) and was entering in the previous time slot as well ($RC_{l,t} = V_{l,t}$).

For water at temperature level l leaving the pipe ($X^D_{l,t} = 1$) we need to ensure that its right coordinate $RC_{l,t}$ is at the most right place and, thus, equal to the pipe volume v:

$$RC_{l,t} \geq v \cdot X^D_{l,t} \qquad \forall l \in S_l, \forall t \in S_{t<n_t}. \tag{4.25}$$

4.5 Overview of hybrid time grid model formulation

Combining all equations presented in this chapter can replace the heat balance without thermal dynamics (2.5) in the unit commitment problem presented in Section 2.1.1. This results in a MILP formulation for the scheduling of CHPs considering heating grid dynamics. This resulting model can be solved directly with off-the-shelf MILP solvers and an overview of this model is given in Model 4.1.

Model 4.1: Hybrid discrete-continuous time grid model

Objective function

$$\min \sum_{t \in S_t} \left(price^{el,buy}_t \cdot p^{buy}_t - price^{el,sell}_t \cdot p^{sell}_t \right) + \sum_{t \in S_t} \sum_{i \in S_g} \left(C^{const}_i \cdot n_{i,t} + C^{el,var}_i \cdot p_{i,t} + C^{heat,var}_i \cdot q_{i,t} \right)$$

Hybrid time grid

$$t_{t+1} = t_t + L_t \qquad \forall t \in S_{t<n_t}$$

$$t_t = tfx_t \qquad \forall t \in S^{fx}_t$$

Supply node

$$n_{i,t} \cdot P^L_i \leq p_{i,t} \leq n_{i,t} \cdot P^U_i \qquad \forall i \in S_g, \forall t \in S_t$$

$$n_{i,t} \cdot Q^L_i \leq q_{i,t} \leq n_{i,t} \cdot Q^U_i \qquad \forall i \in S_g, \forall t \in S_t$$

$$p^{demand}_t = p_t + p^{buy}_t - p^{sell}_t \qquad \forall t \in S_t$$

$$q_t = \sum_l F^C_{l,t} \cdot \rho \cdot c_p \cdot \left(T^{supply}_l - T^{return} \right) \qquad \forall t \in S_{t<n_t}$$

$$\sum_l X^C_{l,t} \leq 1 \qquad \forall t \in S_{t<n_t}$$

$$f^L X^C_{l,t} \leq F^C_{l,t} \leq f^U X^C_{l,t} \qquad \forall l \in S_l, \forall t \in S_{t<n_t}$$

Feasible region of CHP with extraction condensing turbine:

$$\frac{p_t}{p^U} \le \alpha_1 \cdot L_t - \beta_1 \frac{q_t}{Q^U} \qquad \forall t \in S_t$$

$$\frac{p_t}{p^U} \ge \alpha_2 \cdot L_t - \beta_2 \frac{q_t}{Q^U} \qquad \forall t \in S_t$$

$$\frac{p_t}{p^U} \ge -\alpha_3 \cdot L_t + \beta_3 \frac{q_t}{Q^U} \qquad \forall t \in S_t$$

Demand node

$$\sum_l X^D_{l,t} \le 1 \qquad \forall t \in S_{t<n_t}$$

$$f^L X^D_{l,t} \le F^D_{l,t} \le f^U X^D_{l,t} \qquad \forall l \in S_l, \forall t \in S_{t<n_t}$$

$$\sum_l \frac{F^D_{l,t}}{\dot{m}^U/\rho} \le L_t \le \sum_l \frac{F^D_{l,t}}{\dot{m}^L/\rho} + 1 - \sum_l X^D_{l,t} \qquad \forall t \in S_{t<n_t}$$

$$q_t^{demand} \cdot L_t = \sum_l F^D_{l,t} \cdot \rho \cdot c_p \cdot \left(T_l^{supply} - T^{return}\right) \qquad \forall t \in S_{t<n_t}$$

Pipe representation

$$V_{l,t} = v_l^0|_{t=1} + \left(V_{l,t-1} + F^C_{l,t-1} - F^D_{l,t-1}\right)|_{t>1} \qquad \forall l \in S_l, \forall t \in S_t$$

$$\sum_l V_{l,t} = v \qquad \forall t \in S_t$$

$$LC_{l,t} \le v \qquad \forall l \in S_l, \forall t \in S_t$$

$$RC_{l,t} \le v \qquad \forall l \in S_l, \forall t \in S_t$$

$$RC_{l,t} - LC_{l,t} = V_{l,t} \qquad \forall l \in S_l, \forall t \in S_t$$

$$LC^{end}_{l,t} = LC_{l,t} + \sum_{l' \ne l} F^C_{l',t} \qquad \forall l \in S_l, \forall t \in S_{t<n_t}$$

$$RC^{end}_{l,t} = RC_{l,t} + \sum_{l' \ne l} F^D_{l',t} \qquad \forall l \in S_l, \forall t \in S_{t<n_t}$$

$$LC_{l,t} = lc_l^0|_{t=1} + \left(LC^{end}_{l,t-1} - ZC_{l,t}\right)|_{t>1} \qquad \forall l \in S_l, \forall t \in S_t,$$

$$RC_{l,t} = rc_l^0|_{t=1} + \left(RC^{end}_{l,t-1} - ZC_{l,t}\right)|_{t>1} \qquad \forall l \in S_l, \forall t \in S_t$$

$$V_{l,t} \le v \cdot X^P_{l,t} \qquad \forall l \in S_l, \forall t \in S_{t>1}$$

$$LC_{l,t} \le v \cdot X^P_{l,t} \qquad \forall l \in S_l, \forall t \in S_{t>1}$$

$$RC_{l,t} \le v \cdot X^P_{l,t} \qquad \forall l \in S_l, \forall t \in S_{t>1}$$

$$RC_{l,t} \ge v \cdot X^D_{l,t} \qquad \forall l \in S_l, \forall t \in S_{t<n_t}$$

$$ZC_{l,t} \leq f^{U} \cdot \left(1 - X^{P}_{l,t}\right) \qquad \forall l \in S_l, \forall t \in S_{t>1}$$

$$LC_{l,t} \leq v \cdot \left(1 - X^{C}_{l,t}\right) \qquad \forall l \in S_l, \forall t \in S_{t<n_t}$$

4.6 Summary of chapter

The model presented in this chapter and published in [MC20] is based on latest findings in pipeline scheduling which are explained in Section 2.3. The supply temperature was discretized to adapt these scheduling models, that have been designed for scheduling transport of refined petroleum products via pipes, to heating grids. Thus, only a discrete set of supply temperatures is allowed. Of course, this discretization reduces the possible solution space possibly leading to worse results. However, if enough temperature levels are introduced (e.g. every 5 Kelvin), the solution space should remain big enough that this influence remains small. The original pipeline scheduling model uses a continuous time representation [Cas17]. In contrast, our model uses a hybrid discrete-continuous time grid (see Section 4.1) that enables an efficient integration of energy prices and load forecasts that usually vary at discrete time points (e.g. every hour or quarter hour). These two extensions, the discretization of supply temperature and the discrete-continuous time grid, result in an optimization model for heating grids which enables an accurate representation of transport times with floating time points, while staying solvable with off-the-shelf solvers as the resulting problem is a MILP and non-linearities are avoided.

The hybrid discrete-continuous time grid is unique in modeling of heating grids, as approaches in literature usually use a discrete time representation. The pipeline scheduling models presented in [Cas17] allow to consider meshed grids and reversible flows. As the approach presented in this chapter is an extension of these models, the additions allowing meshed grids and reversible flows could be applied for heating grids as well. Consideration of pumping cost and an accurate representation of heat losses will be difficult, as using a volume-centric model formulation there is no explicit variable for mass flow or velocity. However, an integration of approximate heat losses depending on the current supply temperature level is possible by introducing one new parameter per temperature level which represents the reduced supply temperatures at the demand node.

The hybrid time grid approach requires detailed knowledge of grid topology and pipe parameters (e.g. length, diameter, heat loss coefficient). Thus, it seems not suitable for distribution grids with many small pipes, but very suitable for heat transport networks with few large pipes. With the innovative problem formulation it allows an exact calculation of flow duration while resulting in a well to solve MILP problem.

5 Delay matrix approach

One of the major drawbacks of the optimization approaches considering heating grid dynamics discussed in state of the art as well as in the previous chapters is that they all require a detailed grid topology. For real-world use of these models, this leads to a modeling error, because very often not all pipe parameters are known. Furthermore, this requires a high engineering effort, as the heating grid needs to be modeled in detail before an aggregation strategy or another simplification approach allows to build suitable optimization models. In addition, most methods use either an extensive iterative or sequential solution algorithm or strong assumptions (like constant mass flow). Groß uses a different approach by proposing a function f directly linking thermal energy stored in the heating grid q_t^{charge} with current and past supply temperatures of generators T_t^{supply} and the current overall load factor $q_t^{demand} / q^{demand,max}$ [Gro12]:

$$q_t^{charge} = f\left(T_t^{supply}, T_{t-1}^{supply}, T_{t-2}^{supply}, T_{t-3}^{supply}, \dots, \frac{q_t^{demand}}{q^{demand,max}}\right) \qquad \forall t \in S_t. \tag{5.1}$$

Using such a function as in (5.1) enables building a very fast and efficient optimization model if added to a unit commitment and dispatch model without consideration of grid dynamics (Section 2.1.1). In [Gro12] this function is built using a linear regression trained with simulated data. Hence, the engineering effort needed to build a detailed heating grid simulation model remains. Using linear regression mainly shifts computational effort from optimization runs to optimization model parameterization.

In the following, a similar function as (5.1) is derived. However, to reduce engineering efforts for real-world applications, a more direct way to find this function is proposed. Section 5.1 recalls the storage effect of thermal dynamics. Section 5.2 introduces an outer approximation of the thermal energy stored in the heating grid which leads to the delay matrix approach in Section 5.3. This approach was developed in a master thesis [Hai18] and is published in [MHH19]. It allows to consider thermal dynamics of heating grids with all thermal generation in one place. The integration of thermal losses presented in Section 5.3.1 and a detailed real-world case study using this approach are described in [MKV+19]. Section 5.3.2 presents an extension to multiple CHPs in different locations that was developed in an internship [Lor19].

5.1 Dynamics of a heating grid with multiple loads

To recall the main principles of thermal dynamics and the thermal storage effect in a heating grid we extend Example 2.2 with one producer and one load to Example 5.1 with one producer and two loads following [MHH19].

Example 5.1:

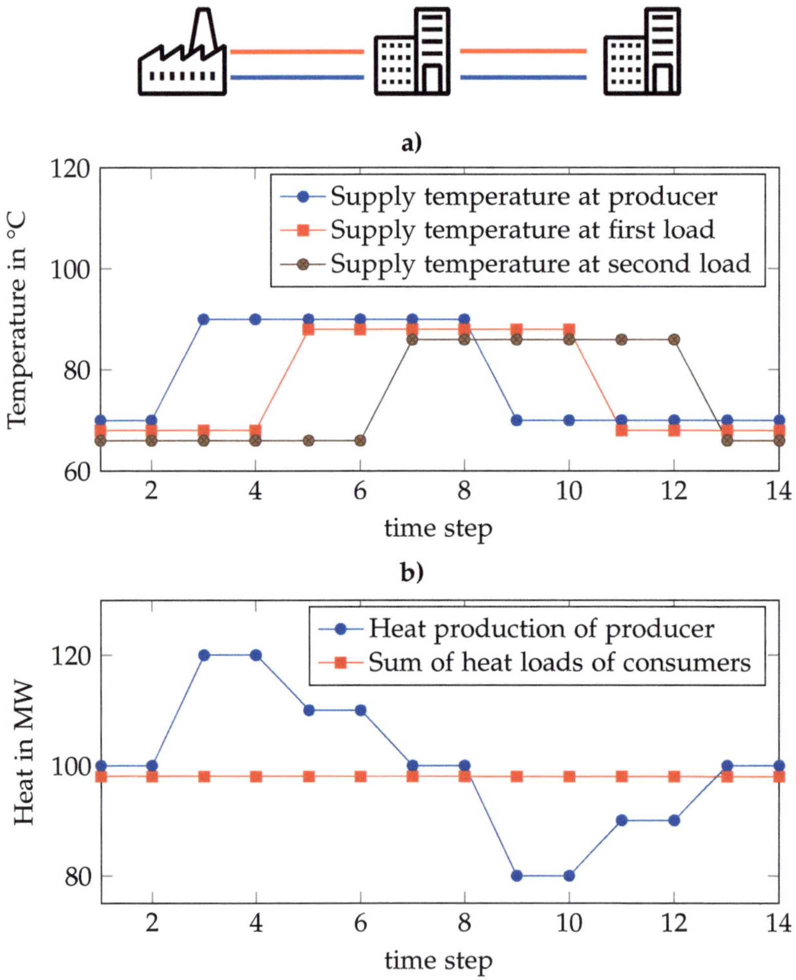

Figure 5.1: Storage behavior of a heating grid with two consumers. a) Supply temperatures at producer and loads. b) Heat production of producer and sum of heat consumption of consumers.

A situation with one producer and two loads is shown in Figure 5.1. Here the increase of supply temperature at the producer in time step 3 reaches the first consumer in time step 5 and the second consumer in time step 7. Thus, the increase of the thermal energy output of the producer starts in time step 3 and is reduced to its original level in two steps. When the increased supply temperature reaches the first consumer in time step 5 this consumer reduces its mass flow. Thus, the overall mass flow is reduced, but the second consumer is still supplied with increased mass flow. This reduction is proportional to the mass flow or load share of consumer one. When the supply temperature increase reaches the second and last consumer, the thermal energy output of the producer is back to its original level and, hence, the charging of the grid ends.
For a decrease of supply temperature at the producer in time step 9 we can observe an inversed behavior (similar as in Example 2.2). The thermal energy output of the producer is reduced immediately leading to a discharge of the grid. The thermal energy production gets back to its original level in two steps. When the first consumer is reached in time step 11, the producer increases its output with the now increased mass flow of consumer one. When the second consumer is reached in time step 13 the thermal energy output of the producer reaches its original level and discharging of the grid ends. Similar to Example 2.2 thermal losses are represented in the reduced supply temperatures at the loads.

5.2 Outer approximation of grid storage behavior

To integrate the thermal storage behavior of a heating grid explained in the previous chapter into a basic scheduling problem for CHPs (see Section 2.1.1), we try to extend the standard models for electric battery storage. Thus, we introduce one non-negative, real variable representing the state of charge SOC_t^{heat} in time slot t and one real variable representing the thermal energy charged or discharged q_t^{charge} in time slot t. Using these variables, the dynamic behavior of the state of charge is described with

$$SOC_{t+1}^{heat} = SOC_t^{heat} + q_t^{charge} \qquad \forall t \in S_t. \tag{5.2}$$

To integrate thermal energy storage into the basic unit commitment problem (Section 2.1.1), we adapt the thermal energy balance (2.5) to

$$\sum_{i \in S_g} q_{i,t} = \sum_{i \in S_d} q_{i,t}^{demand} + q_t^{loss} + q_t^{charge} \qquad \forall t \in S_t. \tag{5.3}$$

For a battery, the state of charge SOC_t^{heat} is limited by the capacity limits of the storage and the energy of charge or discharge q_t^{charge} is limited by the maximum charging and maximum discharging power. For dedicated thermal storage like storage tanks these limits are usually known. However, it is more complicated to derive these limits for the inherent thermal storage of a heating grids. In the following we introduce such limits suitable to approximate the thermal dynamics of a heating grid as we showed in [MHH19]. Looking at Example 2.2 and Example 5.1 there is a clear upper limit on the energy charged to as well as the energy discharged from the heating grid being linked to the maximum possible supply temperature difference $\left(T^U - T_t^{supply}\right)$ in time slot t. With the equation for the change of inner energy (2.9) we can relate this maximum temperature difference to a maximum energy volume charged to or discharged from the grid and get

$$-c_p\dot{m}_t\left(T^U - T_t^{supply}\right) \leq q_t^{charge} \leq c_p\dot{m}_t\left(T^U - T_t^{supply}\right) \qquad \forall t \in S_t \tag{5.4}$$

where T^U is the maximum possible supply temperature, T_t^{supply} is the originally planned (minimum) supply temperature in time slot t, $\dot{m}_t$ is the mass flow from the generation site in time slot t and c_p is the specific heat capacity of water.

The upper bound for the state of charge SOC_t^{heat} of the thermal energy stored in the heating grid in time slot t is more complicated to derive. Assuming we know the maximum transport delay τ^{max} to the last consumer in the heating grid, we can sum up the maximum charging power over this time horizon to find an upper limit for the state of charge SOC_t^{heat} resulting in

$$0 \leq SOC_t^{heat} \leq \sum_{tt=t-\tau^{max}}^{t} c_p\dot{m}_t\left(T^U - T_{tt}^{supply}\right) \qquad \forall t \in S_t. \tag{5.5}$$

However, with this upper bound of the state of charge SOC_t^{heat} we only get an outer approximation of the possible energy stored in the grid. In scenarios with multiple loads not all consumers are served with the maximum transport delay τ^{max}. Accordingly, the thermal energy charged to the heating grid cannot reach the maximum level for the full duration τ^{max}. In Example 5.1 this formulation overestimates the storage capabilities of the heating grid, as in time step 5 the supply temperature increase reaches the first consumer (consuming 50 % of the thermal load) and in the following charging the grid with thermal energy is reduced by half. With the representation above it is assumed that maximum charging is still possible.

Thus, to derive a better representation of the thermal storage behavior of heating grids, the delay matrix approach is presented in the following section as published in [MHH19].

5.3 Delay matrix approach for co-located thermal generation

From Figures 2.4 and 5.1 it seems to be possible to derive a function that directly links supply temperature at the producer T_t^{supply} and thermal energy charged or discharged from the grid q_t^{charge} as in (5.1), if constant mass flow is assumed.

Let us investigate the temperature and thermal energy variations in Example 2.2. The temperature increase from time step 2 to time step 3 is

$$T_{t=3}^{supply} - T_{t=2}^{supply}. \tag{5.6}$$

The mass flow $\dot{m}$ does not change between these time steps, which leads to an increased output of thermal energy of the producer of

$$c_p \dot{m} \left(T_{t=3}^{supply} - T_{t=2}^{supply} \right). \tag{5.7}$$

When the increased supply temperature reaches the consumer in time step 5 after a time delay $\tau = 2$, the thermal output of the producer is reduced by the same amount:

$$c_p \dot{m} \left(T_{t=3}^{supply} - T_{t=2}^{supply} \right) = c_p \dot{m} \left(T_{t=5-\tau}^{supply} - T_{t=4-\tau}^{supply} \right). \tag{5.8}$$

As for the outer-approximation of the storage behavior in Section 5.2 we introduce q_t^{charge} as thermal energy being charged to or discharged from the grid resulting in the energy balance

$$\sum_{i \in S_d} q_{i,t}^{demand} + q_t^{loss} + q_t^{charge} = \sum_{i \in S_g} q_{i,t} \qquad \forall t \in S_t. \tag{5.9}$$

In Example 2.2 with a constant head load, changes in the heat output of the producer directly influence the thermal energy stored in the grid q_t^{charge}. Hence, with (5.7) and (5.8) we get

$$\begin{aligned} q_t^{charge} = q_{t-1}^{charge} &+ c_p \dot{m} \left(T_t^{supply} - T_{t-1}^{supply} \right) \\ &- c_p \dot{m} \left(T_{t-\tau}^{supply} - T_{t-\tau-1}^{supply} \right) \qquad \forall t \in S_{t>1}, \end{aligned} \tag{5.10}$$

where T_t^{supply} is the supply temperature at the producer in time step t, τ is the transport time from producer to consumer and $\dot{m}$ is the constant mass flow. For time varying mass flow $\dot{m}_t$ considered to be a parameter, (5.10) becomes

$$\begin{aligned} q_t^{charge} = q_{t-1}^{charge} &+ c_p \left(\dot{m}_t T_t^{supply} - \dot{m}_{t-1} T_{t-1}^{supply} \right) \\ &- c_p \left(\dot{m}_{t-\tau} T_{t-\tau}^{supply} + \dot{m}_{t-\tau-1} T_{t-\tau-1}^{supply} \right) \qquad \forall t \in S_{t>1}. \end{aligned} \tag{5.11}$$

However, the time varying transport delay coming with time varying mass flow is not represented correctly in the above equation as there is only one transport delay τ possible. To cope with this problem, we introduce a matrix M allowing flexible transport delays. Every row l in this matrix represents the start of a temperature increase in time step l. This allows to have different transport delays throughout the optimization horizon in line with the varying mass flows $\dot{m}_t$. Rows l in matrix M have non-zero

Table 5.1: Delay matrix $M[l,t]$ for Example 5.1

l \ t	1	2	3	4	5	6	7
1	0	0	0.5	0	0.5	0	0
2	0	0	0	0.5	0	0.5	0
3	0	0	0	0	0.5	0	0.5
4	0	0	0	0	0	0.5	0
5	0	0	0	0	0	0	0.5

entries in columns t only if an increase of consumption in time step l reaches a consumer in time step t. The value of this non-zero entry is the share of consumption of the respective consumer.

In Table 5.1 the matrix for Example 5.1 is shown. As the transport delay to the first consumer is two time steps, the increase of supply temperature at the producer in time step $l = 3$ would reach the first consumer in time step $t = 5$. The second consumer is reached after a transport delay of 4 time steps. Thus, the increase of supply temperature at the first consumer in time step $l = 3$ reaches the second consumer in time step $t = 7$. As both consumers are having equal load share, the non-zero entries of delay matrix M in row 3 are 0.5 in column 5 and 0.5 in column 7. For all other time steps l the rows are filled accordingly. If the transport delay is not an integer number of time steps, the consumption share can be split up between two or more columns to achieve a better representation of the real-world delay.

Adjusting the calculation of q_t^{charge} in (5.11) to integrate delay matrix M we get

$$\begin{aligned} q_t^{charge} &= q_{t-1}^{charge} + c_p\left(\dot{m}_t T_t^{supply} - \dot{m}_{t-1} T_{t-1}^{supply}\right) \\ &- c_p\left(\sum_{l=1}^{n_t} M[l,t]\cdot \dot{m}_l \cdot T_l^{supply} - \sum_{l=1}^{n_t} M[l,t]\cdot \dot{m}_{l-1} T_{l-1}^{supply}\right) \qquad \forall t \in S_{t>1}. \end{aligned} \tag{5.12}$$

As an alternative to this recursive calculation of q_t^{charge}, that we introduced in [MHH19], an explicit formulation was developed in an internship [Lor19]. The energy charged to or discharged from a pipe is the difference between the thermal energy flowing into q_t^{in} and out of q_t^{out} the pipe:

$$q_t^{charge} = q_t^{in} - q_t^{out} \qquad \forall t \in S_t. \tag{5.13}$$

As the thermal energy feeding the pipe is explained with the inner energy balance in (2.9), we get

$$q_t^{in} = c_p \dot{m}_t \left(T_t^{supply} - T^{return}\right) \qquad \forall t \in S_t. \tag{5.14}$$

The thermal energy flowing out of the pipe is the inflowing thermal energy delayed with transport time τ_t:

$$q_t^{out} = q_{t-\tau_t}^{in} = c_p \dot{m}_{t-\tau_t} \left(T_{t-\tau_t}^{supply} - T^{return} \right) \qquad \forall t \in S_t. \tag{5.15}$$

Hence, the energy charged into a pipe q_t^{charge} can be calculated with

$$q_t^{charge} = c_p \left(\dot{m}_t \cdot T_t^{supply} - \dot{m}_{t-\tau_t} \cdot T_{t-\tau}^{supply} \right) \qquad \forall t \in S_t. \tag{5.16}$$

Thus, using delay matrix $M[l,t]$ a similar formulation as in (5.12) is possible:

$$q_t^{charge} = c_p \left(\dot{m}_t \cdot T_t^{supply} - \sum_{l=1}^{n_t} M\left[l,t\right] \cdot \dot{m}_l \cdot T_l^{supply} \right) \qquad \forall t \in S_t. \tag{5.17}$$

(5.12) or (5.17) in combination with (5.9) and delay matrix M (see Table 5.1 for Example 5.1) give a direct link between supply temperature at the producer T_t^{supply} and thermal energy charged to or discharged from the heating grid q_t^{charge}. Thus, they can extend the optimization problem for production scheduling without consideration of grid dynamics in Section 2.1.1 to an optimization problem considering the storage effect induced by supply temperature changes. The consumption share of and transport delay to consumers as well as the mass flow at the producer must be known for this formulation. Model 5.1 shows an overview of this optimization problem at the end of the next section as it includes thermal losses introduced in this section, too.

Remark: As many models in literature, this approach assumes fixed, given transport delays. It does not account for the change in transport delay when a consumer reduces its mass flow with an increased supply temperature arriving. Hence, there is a model error being larger for longer and higher temperature increases. This model error will reduce if variable mass flows are introduced. The approach presented in Section 5.3.2 to integrate a second CHP can be a starting point for this extension as here two different mass flows are considered.

5.3.1 Integration of thermal losses

To consider the additional heat losses caused by the increase of supply temperature, we introduced the following approach in [MKV+19]. In real-world heating grids, there are usually only a few thermal generation points which are measured with high sampling rate and hundreds of loads normally not measured with this accuracy. Thus, thermal demand forecasts with high time resolution are usually predicted based on measurements of past generation at different ambient conditions. Accordingly, those load forecasts $q_{i,t}^{demand}$ will include the thermal losses of the heating grid with the typical supply temperature control. However, increasing the supply temperature to store thermal energy within the heating grid increases thermal losses. This should be considered as q_t^{loss} in (5.9). Recall (2.13) describing the heat losses of a pipe filled with a liquid at one temperature:

$$q^{loss} = k_V \pi d L \left(T^{inside} - T^{amb} \right).$$

As ambient temperature T^{amb} and pipe parameters do not change with an increase of temperature, the increased losses are proportional to the temperature inside the pipe T^{inside}. Thus, we can introduce a loss factor α_{loss} and using a time discrete formulation we get

$$q_t^{loss} = \alpha_{loss} \cdot T_t^{inside}. \tag{5.18}$$

Now the question remains, how to best replace T_t^{inside} with a formulation using the supply temperature at the producer T_t^{supply}. Heat losses first lead to a reduction of temperature of the transport medium. This does not effect the thermal energy output of the generation until this lowered temperature arrives at the consumer and the consumer increases its consumed mass flow. Using delay matrix M we can describe this delayed arrival by

$$q_t^{loss} = \alpha_{loss} \cdot \sum_{l=1}^{n_t} M[l,t] \, T_l^{supply} \qquad \forall t \in S_t. \tag{5.19}$$

In comparison to the dynamic thermal loss calculation in (2.15) this equation neither depends on mass flow $\dot{m}_t$ nor on transport time τ_t. If T_l^{supply} is the pipe inflow temperature, it overestimates the thermal losses as in reality the temperature drops along the pipe. However, for typical operation scenarios this linearized loss calculation should represent the thermal losses sufficiently well, as transport time τ_t has a minor influence on thermal losses in comparison to inflow and ambient temperature. The influence of transport time τ_t in (2.15) grows only for very slow velocity and thus high transport times. Figure 5.2 shows the thermal losses for different pipe diameters and different supply temperatures at different typical velocities in heating grids as calculated from (2.15). It illustrates the behavior of thermal losses described above, as the velocity is proportional to the mass flow and inverse to the transport time.

Model 5.1 summarizes the resulting model formulation for the delay matrix approach considering additional thermal losses induced by the increased supply temperature. These additional thermal losses should be considered in the operations planning problem, especially if a distribution heating grid with a high number of pipes with small diameters is optimized. For transmission grids, which have only pipes with large diameters, thermal losses are often insignificant.

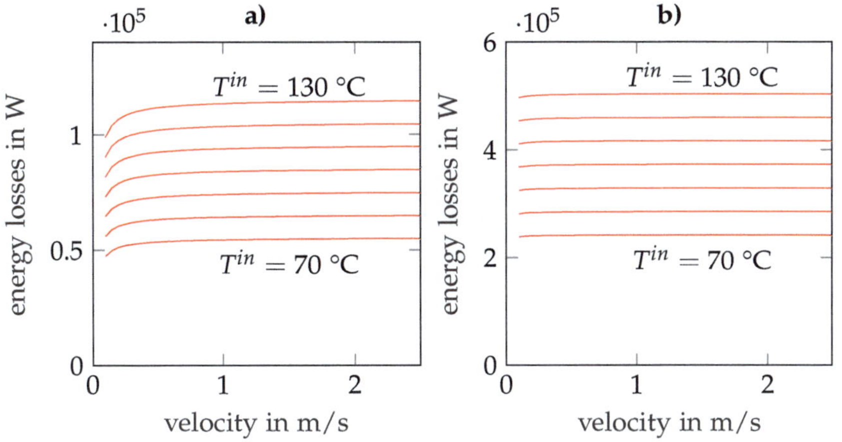

Figure 5.2: Thermal losses in a pipe with 5 km length and a) 0.1 m or b) 0.7 m diameter depending on flow velocity for different inflow temperature levels in 10 K steps from 70 °C (lowest) to 130 °C (highest)

Model 5.1: Delay matrix model for co-located CHPs with thermal losses

Objective function

$$\min \sum_{t \in S_t} \left(price_t^{el,buy} \cdot p_t^{buy} - price_t^{el,sell} \cdot p_t^{sell} \right) + \sum_{t \in S_t} \sum_{i \in S_g} \left(C_i^{const} \cdot n_{i,t} + C_i^{el,var} \cdot p_{i,t} + C_i^{heat,var} \cdot q_{i,t} \right)$$

Inequality Constraints

$$n_{i,t} \cdot P_i^L \leq p_{i,t} \leq n_{i,t} \cdot P_i^U \qquad \forall i \in S_g, \forall t \in S_t$$

$$n_{i,t} \cdot Q_i^L \leq q_{i,t} \leq n_{i,t} \cdot Q_i^U \qquad \forall i \in S_g, \forall t \in S_t$$

Equality Constraints

$$p_t^{demand} = \sum_{i \in S_g} p_{i,t} + p_t^{buy} - p_t^{sell} \qquad \forall t \in S_t$$

$$\sum_{i \in S_d} q_{i,t}^{demand} = \sum_{i \in S_g} q_{i,t} - q_t^{loss} - q_t^{charge} \qquad \forall t \in S_t$$

$$q_t^{charge} = c_p \left(\dot{m}_t \cdot T_t^{supply} - \sum_{l=1}^{n_t} M[l,t] \cdot \dot{m}_l \cdot T_l^{supply} \right) \qquad \forall t \in S_t$$

$$q_t^{loss} = \alpha_{loss} \cdot \sum_{l=1}^{n_t} M[l,t] \, T_l^{supply} \qquad \forall t \in S_t$$

The constraints for the feasible regions of the CHPs need to be added matching their type (e.g. (2.3) for CHPs with extraction condensing turbine).

5.3.2 Integration of distributed thermal generation

To reduce thermal losses, heating grids are usually supplied at minimum possible temperature. As there is a limitation on velocity and flow in a pipe, this minimum temperature can be calculated based on (2.9). To enable integration of distributed thermal generation into the delay matrix approach assume in the following that all CHPs will always feed at minimum temperature and, thus, maximum flow.

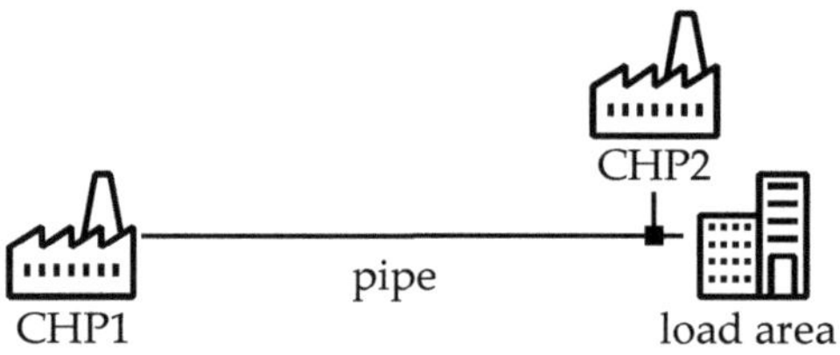

Figure 5.3: A heating grid with a second smaller CHP

To introduce additional CHPs, first, we look at the situation shown in Figure 5.3. Here, one long pipe is connecting one large CHP1 with a load area. A second smaller CHP2 is directly located at the load area feeding the load area without important transport delay. Assuming we always supply with minimum supply temperature and thus maximum flow, we get two possible flows in the pipe in every time slot: One higher flow if CHP2 is turned off and one slightly lower flow if CHP2 is running.

If CHP2 is switched on ($n_{2,t} = 1$), it provides its maximum electric P_2^U and heat Q_2^U output:

$$p_{2,t} = \begin{cases} 0 & n_{2,t} = 0 \\ P_2^U & n_{2,t} = 1 \end{cases} \quad \forall t \in S_t, \tag{5.20}$$

$$q_{2,t} = \begin{cases} 0 & n_{2,t} = 0 \\ Q_2^U & n_{2,t} = 1 \end{cases} \quad \forall t \in S_t. \tag{5.21}$$

At times when the thermal energy storage of the grid is not used (times of constant supply temperature), the thermal output of CHP1 $q_{1,t}$ is reduced accordingly, leading to

$$q_{1,t} = \begin{cases} q_t^{demand} + q_t^{loss} & n_{2,t} = 0 \\ q_t^{demand} + q_t^{loss} - q_{2,t} & n_{2,t} = 1 \end{cases} \quad \forall t \in S_t. \tag{5.22}$$

Hence, the mass flow $\dot{m}_t$ in the pipes from CHP1 to the load area varies with the commitment status $n_{2,t}$ of CHP2:

$$\dot{m}_t = \begin{cases} \frac{1}{c_p \cdot \left(T_t^{supply} - T^{return}\right)} \left(q_t^{demand} + q_t^{loss}\right) & n_{2,t} = 0 \\ \frac{1}{c_p \cdot \left(T_t^{supply} - T^{return}\right)} \left(q_t^{demand} + q_t^{loss} - q_{2,t}\right) & n_{2,t} = 1 \end{cases} \quad \forall t \in S_t. \tag{5.23}$$

Thus, there is a limited amount of possibilities of temperature propagation in the pipe depending on current and past commitment states of CHP2. Assuming a scenario with a time delay limited to variations between two and three time slots, all possible temperature propagations are shown in Figure 5.4. The running status of CHP2 $n_{2,t}$ in past time slots is encoded with $Comb_i$ as shown in Table 5.2.

Table 5.2: Running status of CHP2 in past time steps depending on $Comb_i$

i	$Comb_i$	$n_{2,t}$	$n_{2,t-1}$	$n_{2,t-2}$
1	000	0	0	0
2	001	0	0	1
3	010	0	1	0
4	011	0	1	1
5	100	1	0	0
6	101	1	0	1
7	110	1	1	0
8	111	1	1	1

For the situation shown in Figure 5.4, matrix $M[l,t]$ has only two entries per row and (5.17) becomes

$$q_t^{charge} = c_p \left(\dot{m}_t \cdot T_t^{supply} - s^1 \cdot \dot{m}_{t-1} \cdot T_{t-1}^{supply} - s^2 \cdot \dot{m}_{t-2} \cdot T_{t-2}^{supply}\right) \quad \forall t \in S_t. \tag{5.24}$$

With lengths d^1 and d^2 defined as illustrated in the top pipe in Figure 5.4, parameters s^1 and s^2 can be calculated with the propagation Δx_t of the transport medium in a time slot t as follows:

$$s^1 = \frac{d^1}{\Delta x_{t-1}}, \tag{5.25}$$

$$s^2 = \frac{d^2}{\Delta x_{t-2}}. \tag{5.26}$$

There is one s_i^1 and one s_i^2 for every combination i of current and past commitment states of CHP2 and thus current and past mass flows. As on the right side of Figure 5.4, the values of s_i^1 and s_i^2 can be calculated for every possible combination i of

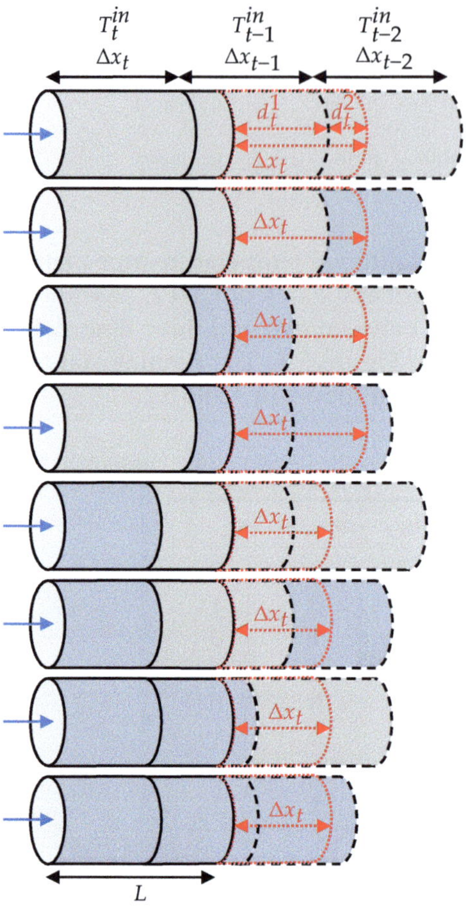

$Comb_i$	s_i^1	s_i^2
000	0.6	0.4
001	0.6	0.48
010	0.52	0.57
011	0.52	0.68
100	0.43	0.4
101	0.43	0.48
110	0.32	0.56
111	0.32	0.68

Figure 5.4: Possibilities of temperature propagation in a pipe with a major CHP1 at the beginning and a smaller CHP2 at the end (close to the load area) for different running states of CHP2 (bright: off, dark: on). The blue arrow indicates the flow direction.

current and past commitment decisions for CHP2 ($Comb_i$). Introducing $y_{t,i}$ as a binary variable being one if and only if case i is active, (5.24) becomes

$$q_t^{charge} = c_p \dot{m}_t \cdot T_t^{supply} - c_p \sum_{i=1}^{8} y_{t,i} \cdot \left(s_i^1 \cdot \dot{m}_{t-1} \cdot T_{t-1}^{supply} + s_i^2 \cdot \dot{m}_{t-2} \cdot T_{t-2}^{supply} \right) \qquad \forall t \in S_t. \quad (5.27)$$

The mass flow $\dot{m}_t$ in the pipe in time slot t depends on the combination of current and past commitment decisions of CHP2, too. As for every case i we know this combination of current and past mass flows, we can introduce new parameters $s_i^{m,1}$ and $s_i^{m,2}$ combining s_i^1 and s_i^2 with the corresponding mass flow:

$$s_i^{m,1} = s_i^1 \cdot \dot{m}_{t-1} \qquad \forall i \in S_c, \quad (5.28)$$

$$s_i^{m,2} = s_i^2 \cdot \dot{m}_{t-2} \qquad \forall i \in S_c. \quad (5.29)$$

S_c denotes the set of all possible combinations to run or not run the CHPs in current and past time slots.

With these parameters, (5.27) becomes

$$q_t^{charge} = c_p \left(\dot{m}_t \cdot T_t^{supply} - \sum_{i=1}^{8} y_{t,i} \cdot \left(s_i^{m,1} \cdot T_{t-1}^{supply} + s_i^{m,2} \cdot T_{t-2}^{supply} \right) \right) \qquad \forall t \in S_t. \quad (5.30)$$

This equation still contains a multiplication of one binary variable ($y_{t,i}$) with one real variable (T_{t-1}^{supply} or T_{t-2}^{supply}). Several MILP solvers, for example Gurobi [Gur16], are able to handle such equations. If the chosen solver does not support this type of equation, bigM constraints or generalized disjunctive programming can be used to formulate this equation as MILP [RG94, GBS+12].

Of course, for every time slot t only one case i representing a combination of current and past mass flows can be selected.

$$\sum_{i=1}^{8} y_{t,i} = 1 \qquad \forall t \in S_t \quad (5.31)$$

The following equations ensure that $y_{t,i} = 1$ only if the correct combination of current and past mass flows is active. Matching Table 5.2 we get

$$
\begin{aligned}
n_{2,t} + n_{2,t-1} + n_{2,t-2} + y_{t,1} &\geq 1 && \forall t \in S_t \\
n_{2,t} + n_{2,t-1} + (1 - n_{2,t-2}) + y_{t,2} &\geq 1 && \forall t \in S_t \\
n_{2,t} + (1 - n_{2,t-1}) + n_{2,t-2} + y_{t,3} &\geq 1 && \forall t \in S_t \\
n_{2,t} + (1 - n_{2,t-1}) + (1 - n_{2,t-2}) + y_{t,4} &\geq 1 && \forall t \in S_t \\
&\vdots \\
(1 - n_{2,t}) + (1 - n_{2,t-1}) + (1 - n_{2,t-2}) + y_{t,8} &\geq 1 && \forall t \in S_t.
\end{aligned}
\tag{5.32}
$$

With $Comb_i$ as shown in the table in Figure 5.4, (5.32) can be reformulated more generally as

$$
\sum_{tt=1}^{3} n_{2,t-tt+1} \cdot (1 - Comb_{i,tt}) + (1 - n_{2,t-tt+1}) \cdot Comb_{i,tt} + y_{t,i} \geq 1, \qquad \forall t \in S_t, \forall i \in S_c. \tag{5.33}
$$

(5.30), (5.31) and (5.33) allow to expand the delay matrix approach with a second CHP if they replace (5.12) or (5.17). Model 5.2 presents the resulting MILP model. If the mass flow varies more importantly with the switching of CHP2 as in the example shown in Figure 5.4, additional parameters s_i^1, s_i^2, s_i^3, ... as well as more cases i are needed. (5.30), (5.31) and (5.33) need to be adapted to match this additional number of cases and parameters. Similar changes are needed if more CHPs are added to the optimization. However, the general concept remains:

- enumerate all possible running state combinations of CHPs ($Comb_i$)
- for each state calculate parameters s_i^1, s_i^2, s_i^3, ...
- adapt (5.30), (5.31) and (5.33) to the increased number of combinations

Model 5.2: Delay matrix model for two CHPs considering thermal losses

Objective function

$$
\min \sum_{t \in S_t} \left(price_t^{el,buy} \cdot p_t^{buy} - price_t^{el,sell} \cdot p_t^{sell} \right) + \sum_{t \in S_t} \sum_{i \in S_g} \left(C_i^{const} \cdot n_{i,t} + C_i^{el,var} \cdot p_{i,t} + C_i^{heat,var} \cdot q_{i,t} \right)
$$

Inequality Constraints

$$
\begin{aligned}
n_{i,t} \cdot P_i^L \leq p_{i,t} \leq n_{i,t} \cdot P_i^U && \forall i \in S_g, \forall t \in S_t \\
n_{i,t} \cdot Q_i^L \leq q_{i,t} \leq n_{i,t} \cdot Q_i^U && \forall i \in S_g, \forall t \in S_t
\end{aligned}
$$

Equality Constraints

$$p_t^{demand} = \sum_{i \in S_g} p_{i,t} + p_t^{buy} - p_t^{sell} \qquad \forall t \in S_t$$

$$\sum_{i \in S_d} q_{i,t}^{demand} = \sum_{i \in S_g} q_{i,t} - q_t^{loss} - q_t^{charge} \qquad \forall t \in S_t$$

$$q_t^{loss} = \alpha_{loss} \cdot \sum_{l=1}^{n_t} M\left[l, t\right] T_l^{supply} \qquad \forall t \in S_t$$

$$q_t^{charge} = c_p \left(\dot{m}_t \cdot T_t^{supply} - \sum_{i=1}^{8} y_{t,i} \cdot \left(s_i^{m,1} \cdot T_{t-1}^{supply} + s_i^{m,2} \cdot T_{t-2}^{supply} \right) \right) \qquad \forall t \in S_t$$

$$\sum_{i=1}^{8} y_{t,i} = 1 \qquad \forall t \in S_t$$

$$\sum_{tt=1}^{3} n_{2,t-tt+1} \cdot (1 - Comb_{i,tt}) + (1 - n_{2,t-tt+1}) \cdot Comb_{i,tt} + y_{t,i} \geq 1$$

$$\forall t \in S_t, \forall i \in S_c$$

The constraints for the feasible regions of the CHPs need to be added matching their type (e.g. (2.3) for CHPs with extraction condensing turbine).

5.4 Summary of chapter

This chapter presented the delay matrix approach, which offers a compact MILP formulation and, thus, promises fast optimization runs. Section 5.1 recalled the thermal dynamics of a heating grid and Section 5.2 showed how these dynamics are related with batteries and thermal storage tanks. Based on these findings, Section 5.3 presented the delay matrix approach that summarizes the dynamics of the heating grid in a delay matrix M which only requires share of consumption of and transport time to loads as input parameters. Thus, it can easily be parameterized using measurement data which allows applications to real-world heating grids as we showed in [MKV+19]. This is a major benefit in comparison to other optimization models as they usually require data of all pipes in the heating grid for parameterization (see Section 2.1.4 and Section 2.1.5). Section 5.3.1 presented an extension of the delay matrix approach that considers the additional thermal losses induced by storing thermal energy in the heating grid. As the thermal losses mainly depend on the transport medium temperature, a linear loss factor was introduced which can be estimated for real-world grids using the grid topology as in [MKV+19]. As many approaches in the literature, the delay matrix approach assumes constant transport delays. However, multiple transport delays can be considered as shown in Section 5.3.2. Here, the delay matrix approach for CHPs at one site was extended to a case with one major CHP storing thermal energy in the heating grid and one minor CHP at another location which is not storing thermal energy in the heating grid. To integrate this second CHP, different cases for all possible mass flow combinations are introduced in the formulation. This concept can be used to allow a flexible transport delay in the original delay matrix approach or it can be extended to represent more generation sites.

6 Case studies

In this chapter, the modeling and optimization approaches for district heating grids developed in Chapter 3, Chapter 4 and Chapter 5 are applied to different scenarios. In Section 6.1 two small district heating grid cases are introduced and used to compare the proposed optimization approaches as well as one approach from literature in terms of performance and solution quality. The comparison of the global optimization approach to the sequential approach from literature was published in [MLH19] including an evaluation of the different primal problems. In Section 6.2 the delay matrix optimization approach of Chapter 5 is applied to the district heating grid of the city of Kiel, Germany. This detailed real-world case study shows potential benefits for different generation setups and is published in [MKV^{+}19].

6.1 Small heating grid examples

This section introduces to two small example heating grids, one with one CHP and one with two CHPs, which follow [MLH19]. For these two example grids, we present results for all optimization approaches introduced in this thesis and compare them with results of a sequential optimization proposed in literature. A detailed discussion of similarities and differences in performance of the algorithms concludes this section.

6.1.1 Example cases

Both example cases use the same electricity price profile and the same thermal and electric energy consumption profiles. For the electricity market it is assumed that electric energy can be sold and purchased from the day-ahead electricity market with a small premium for purchasing. As winter days are more promising for storing thermal energy in a heating grid, the electricity price profile is taken from EPEX SPOT DE/AT [EPE] for November 15th, 2017, a winter day with typical price variations, and shown in Figure 6.1.

The thermal load profile in both scenarios is derived from a typical thermal demand for a November day of a German city at ambient air temperatures of 5 °C to 10 °C. It is shown in Figure 6.2. For the electric demand profile a scaled standard load profile for households shown in Figure 6.3 is used.

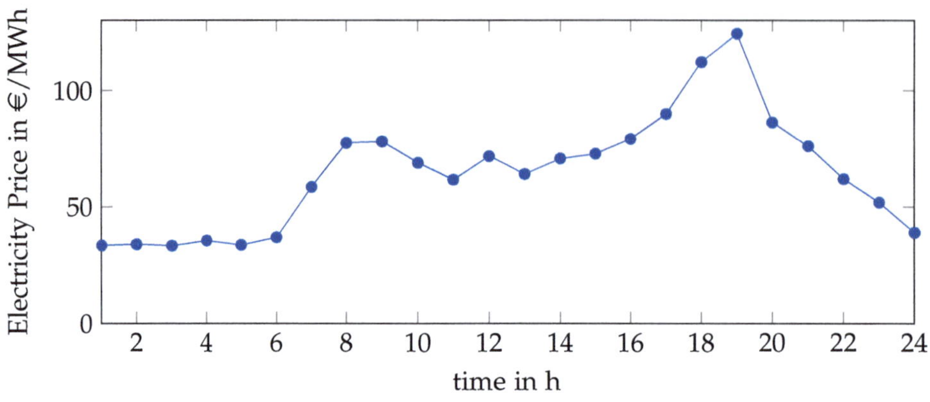

Figure 6.1: Electricity price profile (EPEX SPOT DE/AT, November 15th, 2017). Source: [EPE]

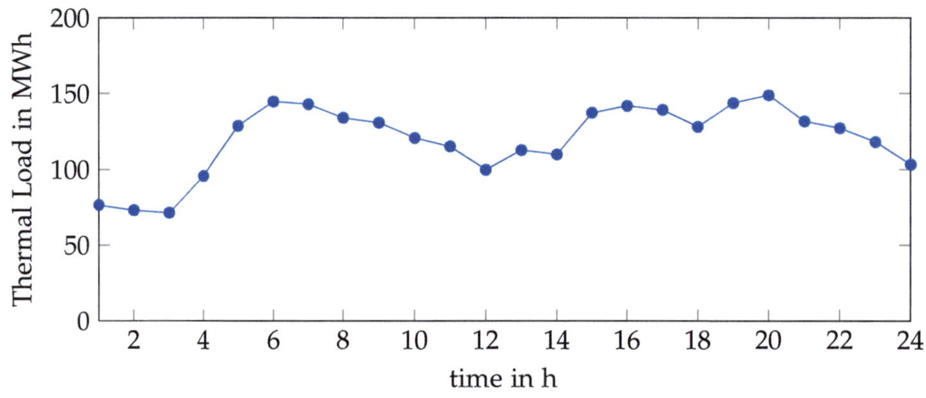

Figure 6.2: Thermal load profile for a November day with ambient air temperatures of 5 °C to 10 °C

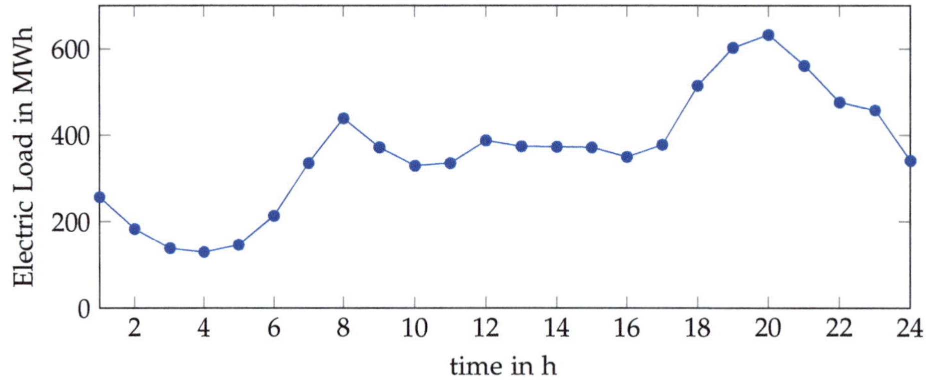

Figure 6.3: Electric load profile (scaled standard load profile)

One CHP case

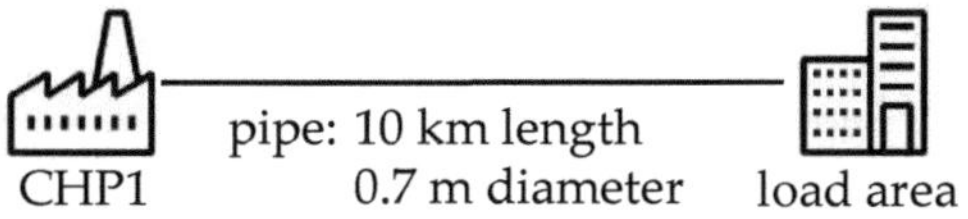

Figure 6.4: Example case with one CHP and one load area

The first scenario considers the most simple heating grid possible: one load area fed by one combined heat-and-power generation plant (CHP1) via a pipe. Pipe dimensions are inspired by the district heating pipeline connecting Mannheim and Heidelberg [KC13] with a length of 10 km (length of Mannheim to Heidelberg connection is 13.5 km) and a diameter of 0.7 m as shown in Figure 6.4. CHP1 is assumed to have a maximum output of 400 MW thermal and 500 MW electric power being generated with an extraction condensing turbine. Thus, the power-to-heat ratio is variable while being limited by the feasible operation region shown in Figure 2.1. CHP1 can heat water to a maximum supply temperature of 130 °C[4], but lower supply temperatures are possible as well. With (2.6) a linear cost function for operating CHP1 is assumed.

Two CHPs case

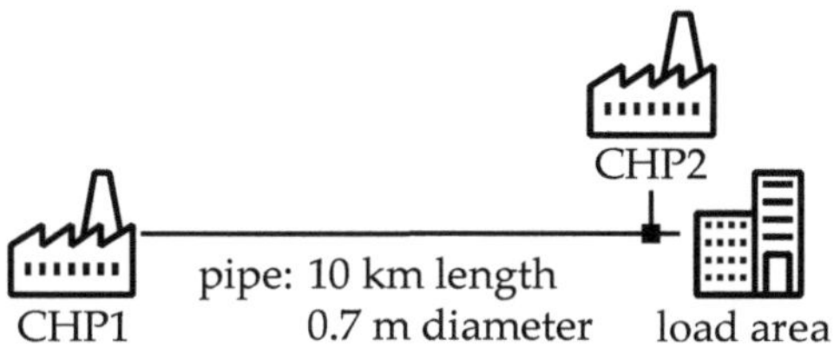

Figure 6.5: Example case with two CHPs and one load area

The second scenario adds a second CHP to the first scenario. This CHP2 is located directly at the load area as shown in Figure 6.5 feeding the load area without a transport delay. CHP2 has typical characteristics of a gas turbine and can produce up to 20 MW of thermal and 20 MW of electric power with a fixed power-to-heat ratio (being 1). It heats water to a constant temperature of 110 °C. This setup is chosen as it is quite common to have one major heat supplier outside the main consumption area and additional smaller heat sources closer to the load area. Gas boilers or gas turbines are common choices for this additional heat sources due to comparably low investment cost, fast start-up and compact size. The characteristics of CHP1, the pipe and the load area remain the same as in the first scenario. A linear cost function is

4 Heating grid pipes are under high pressure. Thus, water inside the pipes is liquid at 130 °C.

assumed for both CHPs. Table 6.1 shows a summary of the parameter differences of the CHPs.

Table 6.1: Main parameters of the CHPs in the example cases

	CHP1	CHP2
Electric output in MW	≤ 500	≤ 20
Thermal output in MW	≤ 400	≤ 20
Supply temperature in °C	≤ 130	$= 110$
Power to heat ratio	see Figure 2.1	constant

6.1.2 Global optimization results with multiparametric delay modeling

In the following, the one CHP and the two CHPs cases presented in the previous section are optimized to global optimality. Multiparametric disaggregation (Section 2.2.2) is used to model bilinear terms and multiparametric delay modeling (Chapter 3) is used to model the variable-dependent time delay in the dual problems leading to the MILP formulation shown in Model 3.1. The primal problems are solved with a local NLP solver with direct implementation of the bilinear terms and using the formulations presented in Section 3.2 to model the variable-dependent time delays. For the primal problem using the finite-volumes model (Model 3.2), 500 equally sized volume elements are used to describe the pipe. Using the primal problem with the hybrid pipe model (Model 3.3), again 500 elements for the remaining flexibility are used. Thermal losses are considered in both the dual as well as the primal problems.

As computational effort for the global optimization approaches is high, all global optimizations were run on an Intel Xeon CPU E5-2650 v3 with 2.3 GHz, 16 cores and 16 GB RAM using the modeling environment Julia/JuMP [Ran16], MILP solver Gurobi 8.1 [Gur16] for the dual problems and NLP solver IPOPT [WB06] to locally solve the primal problems. This case study using the global optimization algorithm is published as part of [MLH19].

One CHP case

The result of the global optimization of the one CHP case using the global optimization scheme with the finite-volumes model as primal problem is shown in Figure 6.6. This result is the solution of the primal problem after 4 iterations having an overall objective value of 291,235.7 €. The lower bound of the dual problem is 288,371.3 € which also provides a lower bound of the original problem. Hence, there is a global optimality gap of 1 %. The result has been achieved in 146.19 s.

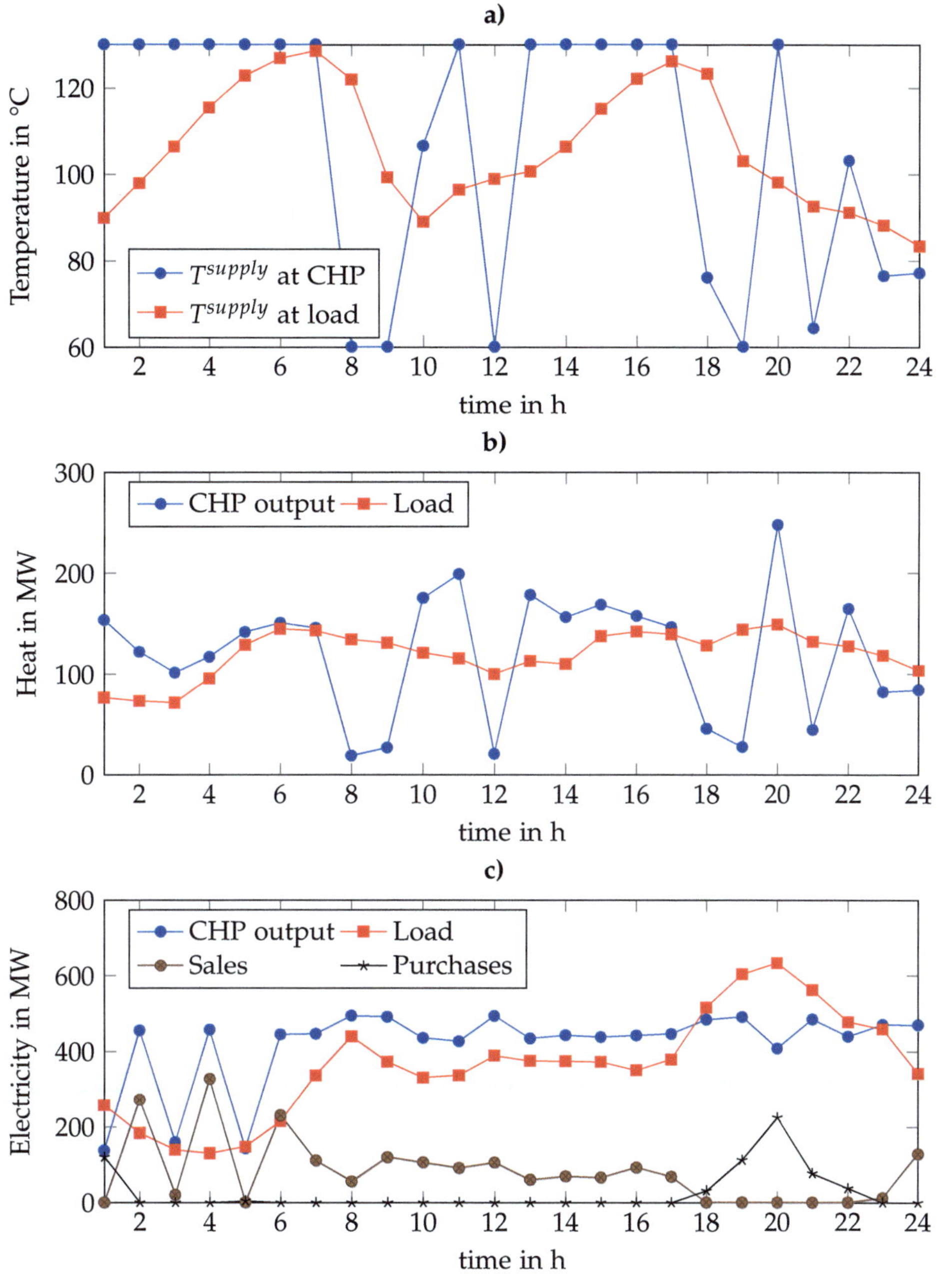

Figure 6.6: Results of global optimization for 1 CHP scenario with finite-volumes model as primal problem. a) Supply temperature at CHP and at load. b) Thermal output of CHP in comparison to thermal load. c) Electric output of CHP, electric load and interactions with electricity market.

Looking at the electric production in Figure 6.6 c), it can be noted that in the early morning hours (hour 0 to hour 6) the electric output of the CHP varies a lot. Here, the price of electricity is approximately at the marginal cost of production. Thus, in one hour it is profitable to produce electricity whereas in the next hour it is not profitable to produce electricity. Running such big ramps in output of a coal fired power plant is usually not wanted. Therefore, in real-world operations one might choose either producing as much as possible or producing as little as possible in this time. Starting from hour 6 the CHP produces as much electricity as possible, as electricity prices are above the marginal cost of production.

In Figure 6.6 a) and Figure 6.6 b) it can be noted that there are major drops in thermal output and supply temperature at hours 8, 12 and 18. From the electricity price profile (Figure 6.1) we note that these are times of higher electricity prices compared to the previous hours. Hence, here the supply temperature at the CHP is lowered to reduce the thermal output of the CHP enabling a higher electric output within the feasible region of the CHP shown in Figure 2.1. As reducing the supply temperature discharges the thermal storage capacity of the heating grid, there needs to be an increase of supply temperature beforehand to store energy within the heating grid. This increase is usually done as late as possible to decrease thermal losses. This can be seen in hours 10 and 11 directly before high-price hour 12, where an important increase of supply temperature charges thermal energy into the heating grid. At the end of the time horizon in hours 21, 23 and 24 after the last price peaks, the supply temperature remains at the lowest level, discharging the thermal energy stored in the heating grid. This is reasonable as the model formulation does not account for thermal energy stored in the heating grid at the end of the time horizon.

However, an unexpected issue can be observed in Figure 6.6 a): The supply temperature is reduced to its minimum level at 60 °C several times within the optimization horizon. This is not feasible in real-world, as mass flow demand would exceed the maximum possible mass flow in a pipe if this temperature arrived at the load. However, with the finite-volumes pipe model these low temperatures at the beginning of the pipe never reach the consumer due to the smoothing effect of the finite-volumes model. Thus, as shown in Figure 6.6 a) the temperature at the load stays within acceptable bounds. Due to the same reason the temperature at the load does not look like a delayed version of the temperature at the CHP in Figure 6.6 a), which it is in real-world.

The hybrid pipe model presented in Section 3.2.2 should allow for a more accurate representation of the temperature propagation. In Figure 6.7, the solution of the global optimization scheme with the hybrid pipe model as primal problem is shown. This result is the feasible solution of the primal problem after 4 iterations having an overall cost of 294,386.0 €. The corresponding global lower bound is 288,265.1 € leading to a global optimality gap of 2.1 %. This result was achieved in 125 s.

In Figure 6.7 a) it can be seen that the supply temperature profile at the load follows

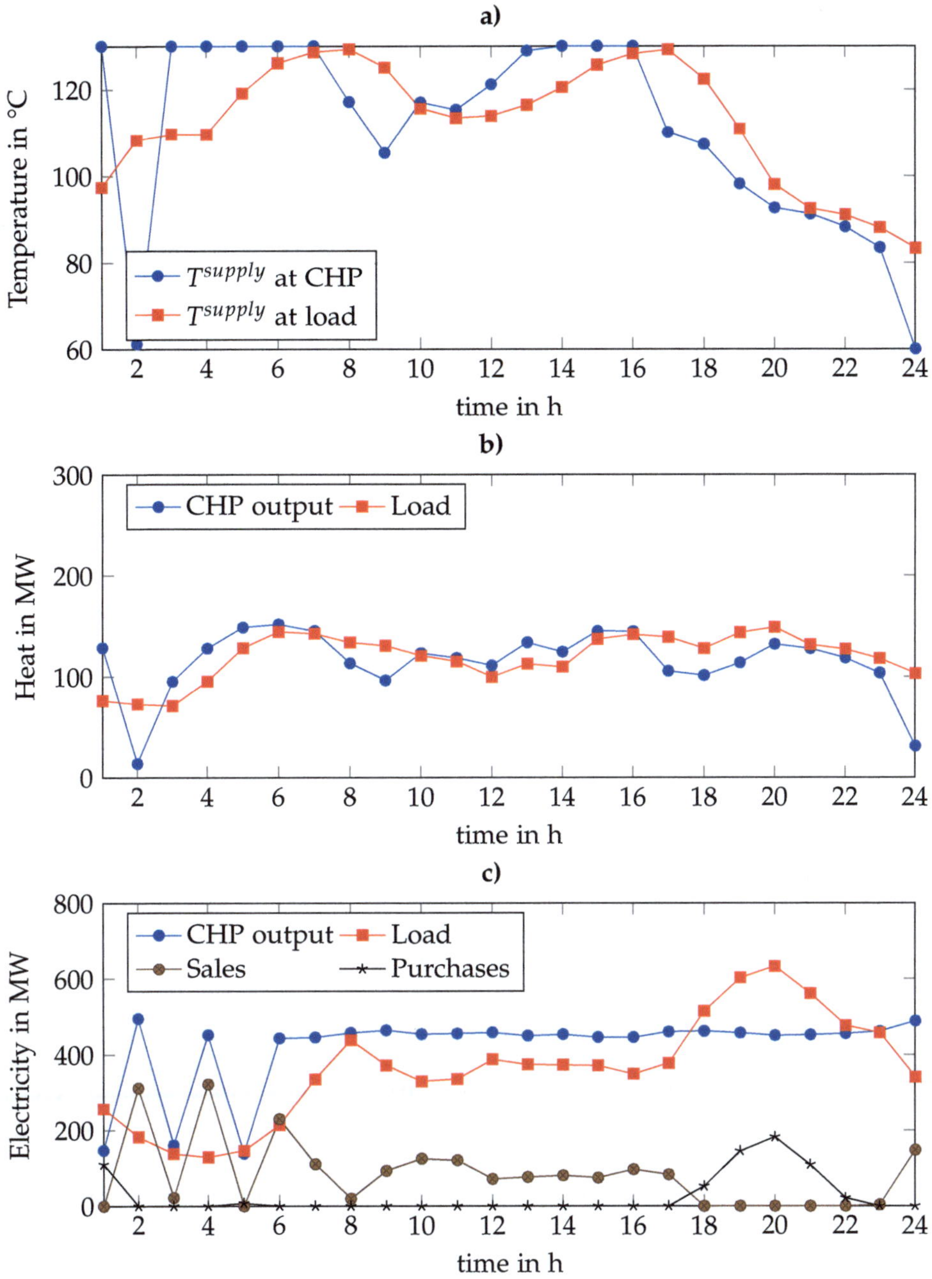

Figure 6.7: Results of global optimization for 1 CHP scenario with hybrid pipe model as primal problem. a) Supply temperature at CHP and at load. b) Thermal output of CHP in comparison to thermal load. c) Electric output of CHP, electric load and interactions with electricity market.

the supply temperature profile at the CHP a lot better than for the finite-volumes pipe model (Figure 6.6 a)). The electric output of the CHP shown in Figure 6.7 c) shows a similar behavior as before with a highly variable profile in the early morning hours up to hour 6 due to the electricity prices swinging around the variable cost of production. After hour 6, the electric output is at its maximum. As shown in Figure 6.7 b) the thermal storage effects are less important than with the previous model. As before, there is some thermal discharge of the heating grid (CHP output below load) around hours 8 and 18, and some charging of the heating grid with thermal energy in the hours before (CHP output above load). However, the amount of thermal energy charged to and discharged from the grid is less important than with the finite-volumes model. This can be explained with the reduced amplitude of the temperature changes at the CHP as smaller temperature changes directly lead to less thermal energy stored in the heating grid.

Two CHPs case

Figure 6.8 shows the result of the global optimization scheme (Chapter 3) using the finite-volumes model as primal problem (Section 3.2.1) for the two CHPs case presented in Section 6.1.1. This result was achieved in 4 iterations and 1,637.0 s. It has an objective value of 348,252.9 € and a lower bound of 330,871.1 €. The global optimality gap is at 4.99 % accordingly. The solution time for solving the dual MILP was limited to 600 s in every iteration to limit the time spent solving one iteration.

The generation of electric energy using CHP1 turns profitable in hour 7. Accordingly, the electric output of CHP1 is increased from minimum to maximum possible level as shown in Figure 6.8 c). CHP2 is switched on in hour 3 and from hours 7 to 21, being the higher price hours. This is reasonable as in contrast to CHP1 for CHP2 a higher thermal output means a higher electric output. Thermal energy is stored in the heating grid (generation of thermal energy above consumption) in the morning hours until hour 6, in hour 11, in hours 13 to 15 and in hour 22 as shown in Figure 6.8 b). In hours 8 to 10, hour 12, hours 18 to 19, hour 21 and starting hour 23 thermal energy is discharged from the heating grid (production of thermal energy below thermal load) with a reduction of supply temperature of CHP1 (see Figure 6.8 a)). As expected, these are hours with local electricity price peaks. As for the one CHP case, due to the smoothing effect of the finite-volumes model, the supply temperature at the load does not follow the supply temperature at CHP1 and the supply temperature at CHP1 reduces to levels infeasible in real-world.

The result of the global optimization of the two CHPs case with the more accurate hybrid pipe model (Section 3.2.2) as primal problem is shown in Figure 6.9. Again with a limitation of the solution time of the dual MILPs to 600 s per iteration, this results was achieved in 6 iterations and 2,907.1 s. It has an objective value of 351,112.0 €, a lower bound of 334,785.9 € and a gap of 4.6 %.

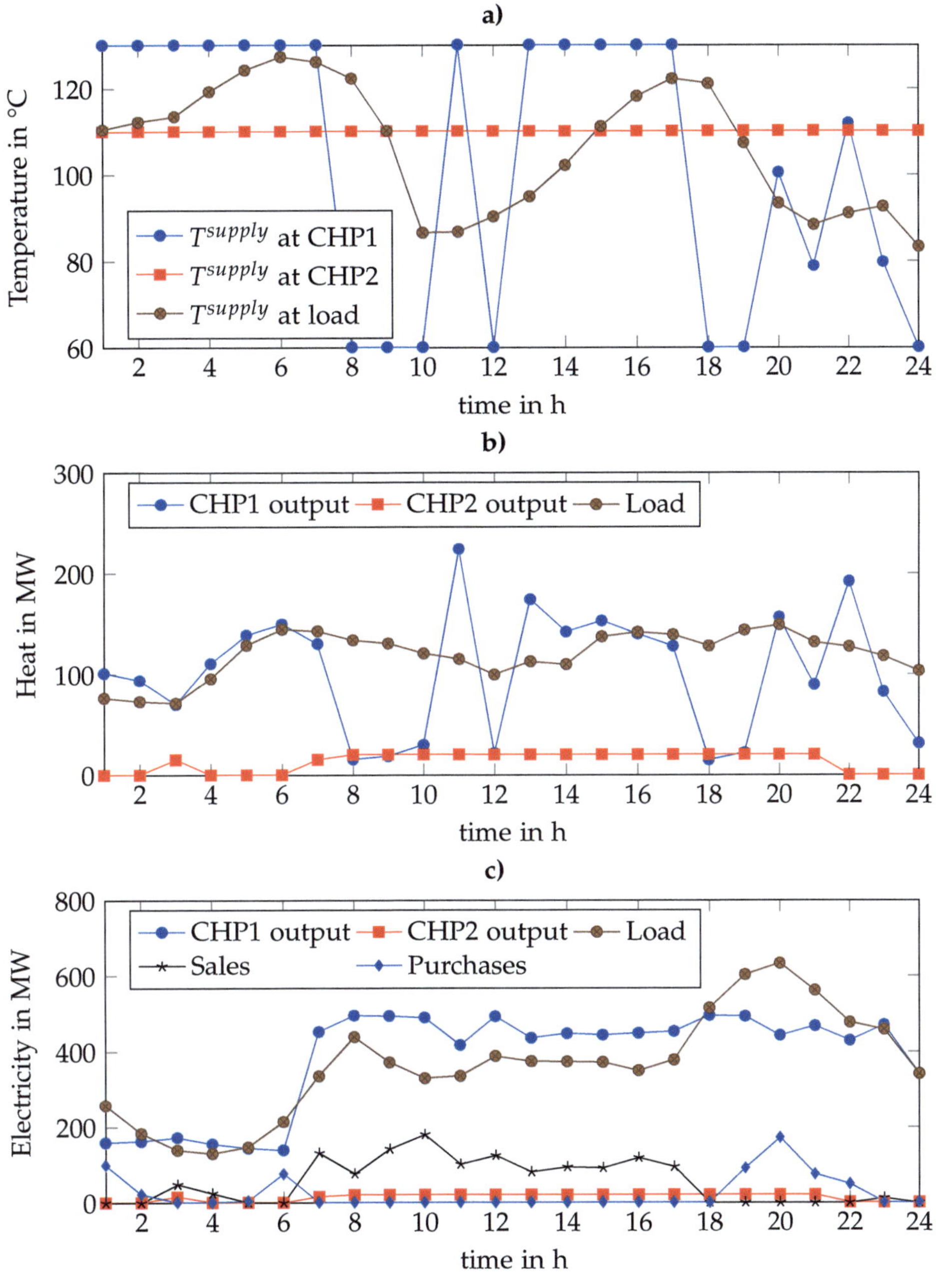

Figure 6.8: Results of global optimization for 2 CHPs scenario with finite-volumes model as primal problem. a) Supply temperature at CHPs and at load. b) Thermal output of CHPs in comparison to thermal load. c) Electric output of CHPs, electric load and interactions with electricity market.

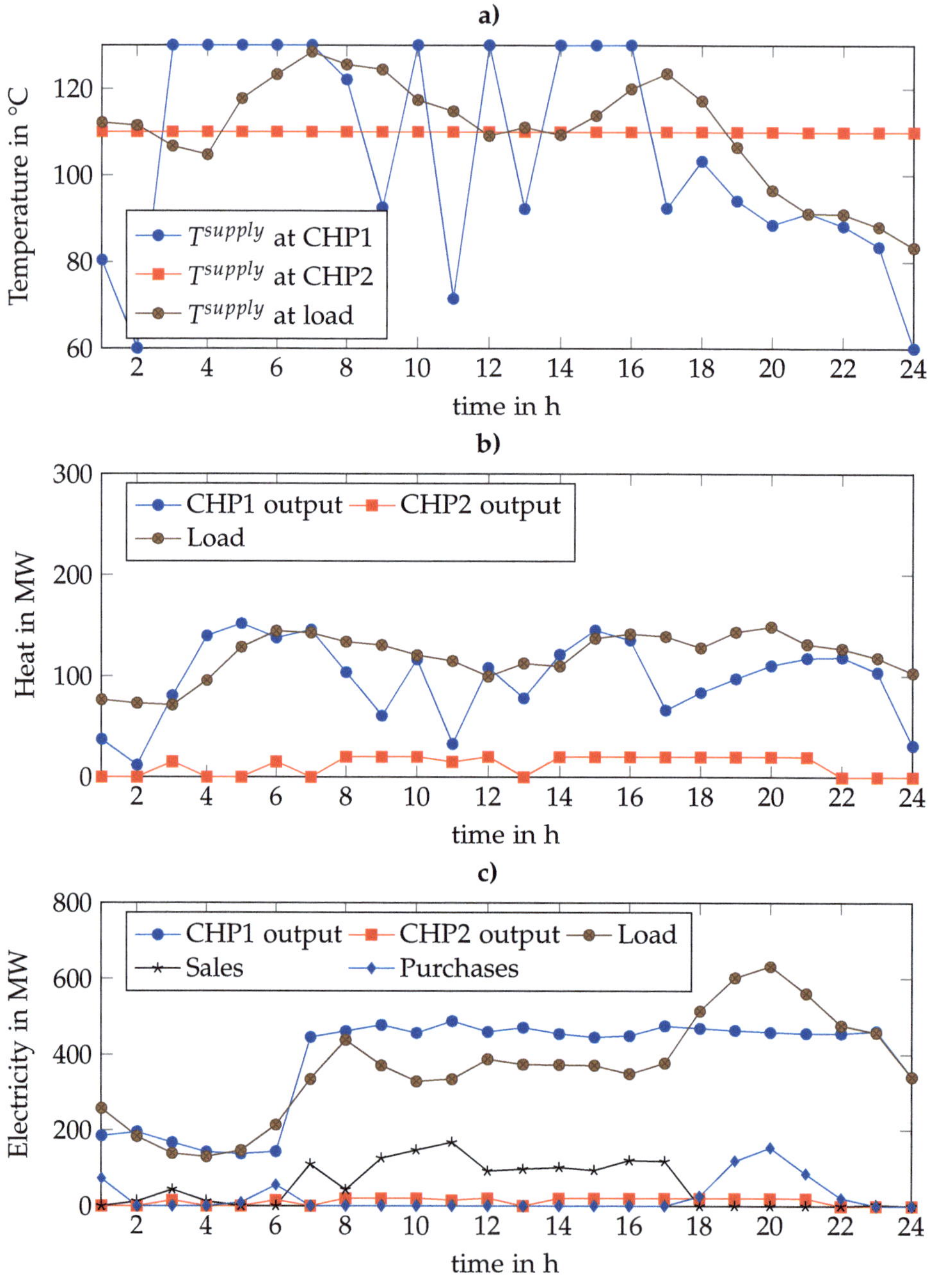

Figure 6.9: Results of global optimization for 2 CHPs scenario with hybrid pipe model as primal problem. a) Supply temperature at CHPs and at load. b) Thermal output of CHPs in comparison to thermal load. c) Electric output of CHPs, electric load and interactions with electricity market.

It can be seen that drops in supply temperature to unreasonably low levels occur less often with the hybrid pipe model as primal problem (Figure 6.9 a)) than with the finite-volumes model (Figure 6.8 a)). However, two large drops remain: one in hour 2 and one in hour 11. The drop to 60 °C in hour 24 is reasonable for the optimization, as this temperature level does not arrive at the load within the optimization horizon. Accordingly, thermal energy is discharged from the heating grid in hour 2 and hour 11. Additional discharging caused by less important and real-world feasible temperature drops occurs in hour 8, hour 13 and starting hour 17 as shown in Figure 6.9 b). Hence, the heating grid is discharged in higher price hour 8 and around high price hour 18. The discharge in hour 11, where the electricity price is lower than the surrounding time slots, seems to be counterintuitive, but might be induced by the starting values of the primal problem, which is only solved to local optimality, as the solver of the dual problem was stopped due to its time limitation. CHP2 runs in hour 3, hour 6, hours 8 to 12 and in hours 14 to 21 being the higher price times. Given the direct positive link between thermal and electric output this is a reasonable choice. As shown in Figure 6.9 c) electric generation of CHP1 turns profitable in hour 7, as the electric output of CHP1 is increased to its maximum at this point in time.

6.1.3 Optimization results with hybrid discrete-continuous time model

The results for the optimization of the two scenarios using the hybrid discrete-continuous time model of Chapter 4 are presented in the following. Heat losses are neglected, as for the studied cases with pipes of diameter 0.7 m they are of minor importance.

These results were achieved on an Intel Silver 4110 CPU with 2.10 GHz and 16 GB RAM using the modeling environment Julia/JuMP [Ran16] and MILP solver Gurobi [Gur16].

One CHP case

Figure 6.10 shows the optimization result using the hybrid time grid model for the one CHP case. To enable a direct comparison with the other optimization methods this result is averaged on hourly bases in Figure 6.11. Allowing eight temperature levels (90 °C, 95 °C, 100 °C, 105 °C, 110 °C, 115 °C, 120 °C and 130 °C) the solution with a MILP gap of 0.0513 % was achieved in 800 s resulting in an overall cost of 292,565.9 €.

In Figure 6.10 c) and Figure 6.11 c) we can observe that the electric output of the CHP in the morning hours from hour 0 to hour 6 shows the same volatility as with the global optimization in Section 6.1.2. This is caused by the same reason: The marginal cost of production is above the electricity price in one hour whereas in the next it

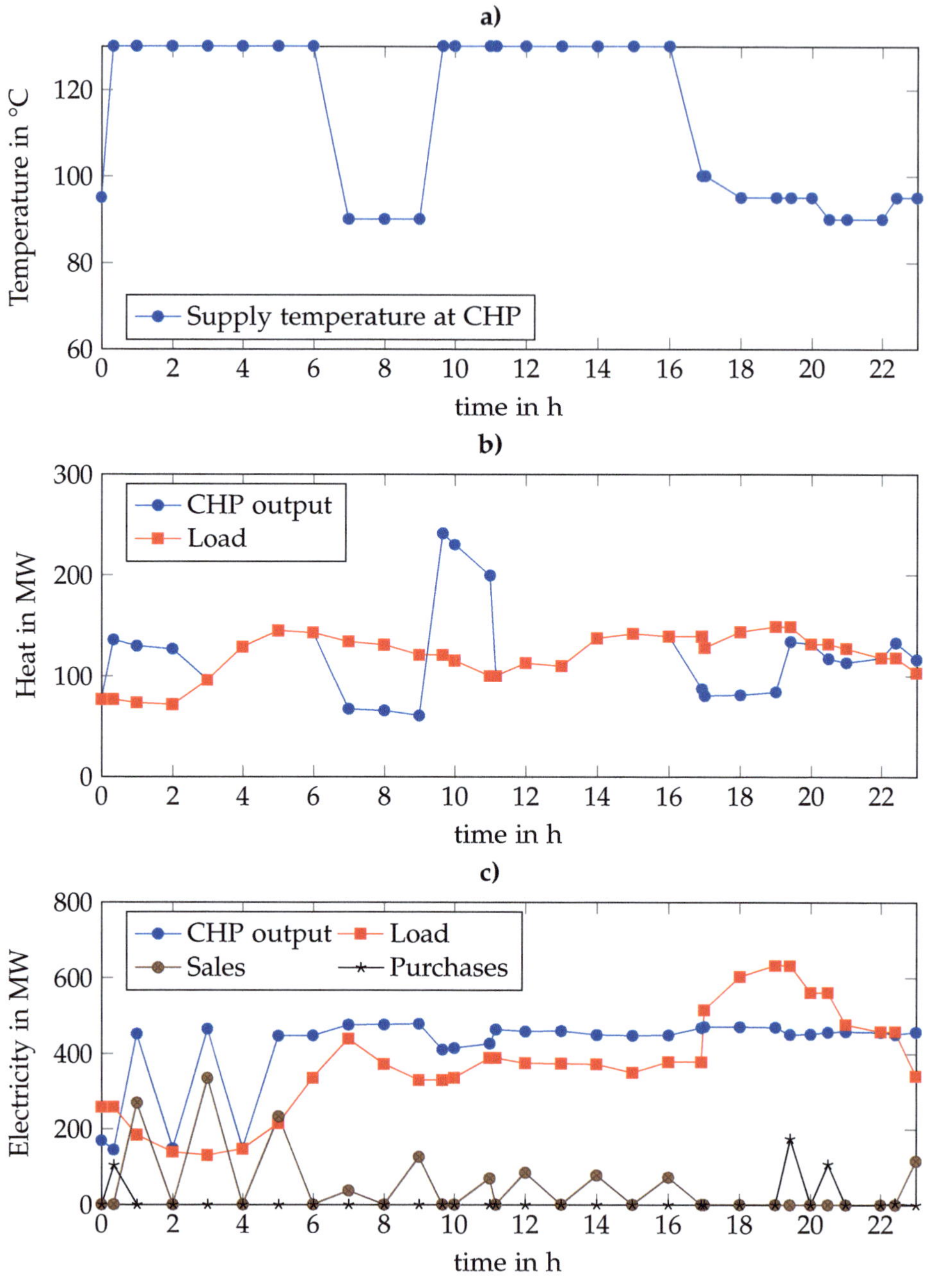

Figure 6.10: Results with exact timing for 1 CHP scenario with hybrid discrete-continuous time model. a) Supply temperature at CHP. b) Thermal output of CHP in comparison to thermal load. c) Electric output of CHP, electric load and interactions with electricity market.

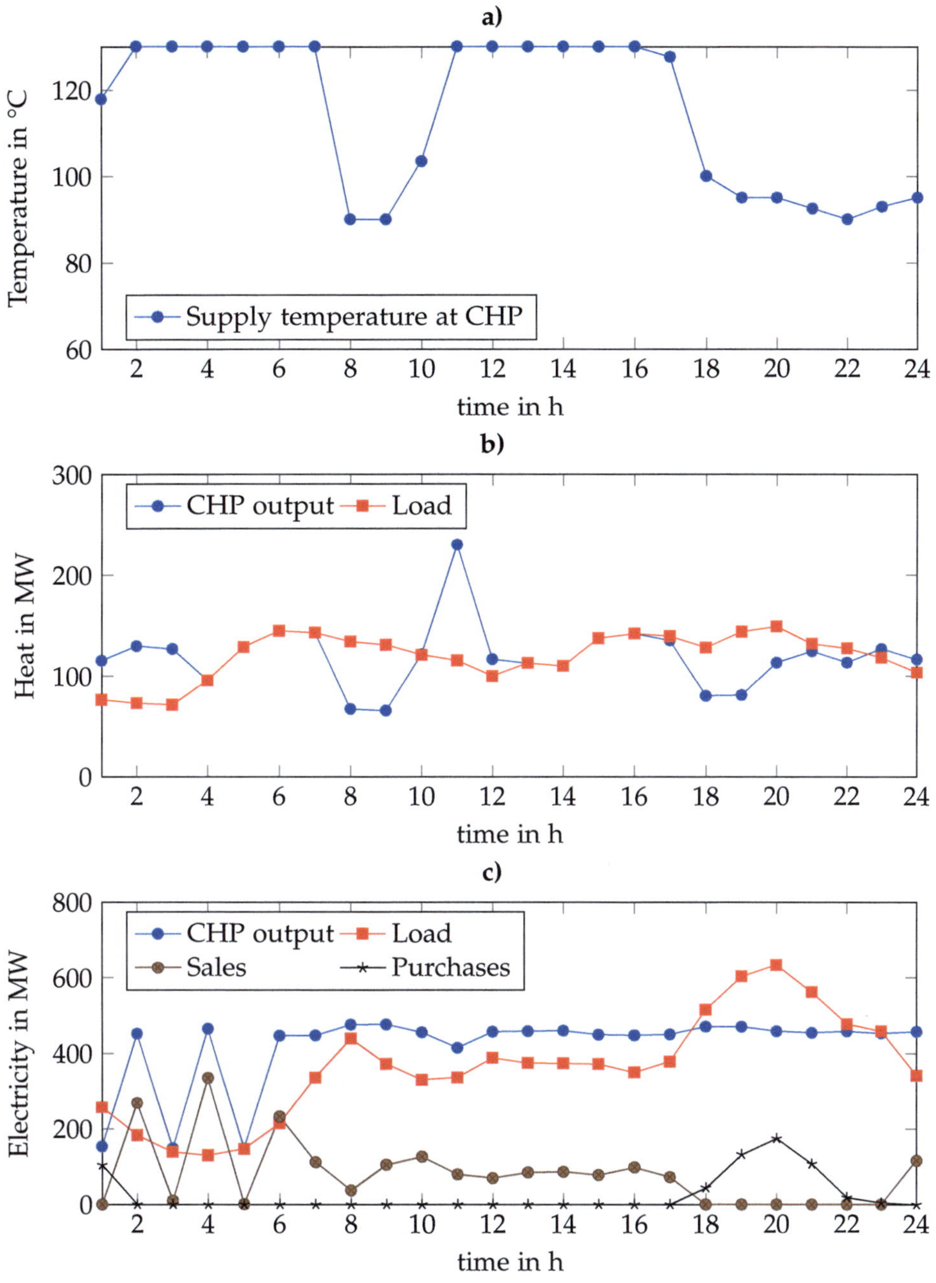

Figure 6.11: Hourly averaged results for 1 CHP scenario with hybrid discrete-continuous time model. a) Supply temperature at CHP. b) Thermal output of CHP in comparison to thermal load. c) Electric output of CHP, electric load and interactions with electricity market.

is below. The CHP runs with maximum possible electric output after hour 6 where generation of electricity becomes profitable for the CHP. Looking at Figures 6.10 a), 6.10 b), 6.11 a) and 6.11 b) the CHP runs with low supply temperatures and thus reduced thermal output around hour 8 and hour 18. As explained in Section 6.1.2, this allows to increase the electric output during these hours of high electricity price. Accordingly, the heating grid is charged with thermal energy by the supply temperature increases in hours 1 to 3 and hour 11 where electricity prices are lower than during the discharging times. This price difference allows to create economic benefits. As thermal losses are not accounted for, the time between charge and discharge of the heating grid is comparably long. With consideration of losses, this duration will reduce as every time slot with high supply temperature leads to increased thermal losses.

To evaluate the accuracy of the hybrid time grid model, we compare the optimization result $q_{1,t}$ of the thermal energy output of the CHP with the result of a simulation $\hat{q}_{1,t}$ of the studied case. This simulation uses the node method (Section 2.1.2) with the optimization result of the supply temperature of the CHP and the thermal load as input. The comparison shows that the thermal generation of the CHP in the optimization result deviates with a root mean square deviation[5] of 29.431 from the simulation result.

Two CHPs case

To integrate CHP2 to the hybrid time grid approach, an energy balance between thermal demand, thermal output of the pipe and thermal generation of CHP2 is added. As before, eight temperature levels are allowed (90 °C, 95 °C, 100 °C, 105 °C, 110 °C, 115 °C, 120 °C and 130 °C). Figure 6.12 shows the optimization result for the scenario with two CHPs being achieved in 800 s with an MILP gap of 0.0554 %. The objective value is 346,490 €. Figure 6.13 shows the same results but with hourly averaged values to enable a direct comparison with the results of the other optimization approaches.

In the morning until hour 6 as well as from hour 11 to hour 17 the supply temperature of CHP1 is increased to store thermal energy in the pipe. Around hour 8 and hours 18 to 22 the supply temperature is set to its minimum possible level, discharging thermal energy from the heating grid. Thus, in these hours the thermal output of CHP1 is reduced and its electric output is increased to sell more or buy less electricity at higher price times. In the morning at hour 7, the operation of both CHPs turns profitable. Electric output of CHP1 is increased to its maximum possible level and CHP2 is switched on. In the evening, at hour 23 CHP2 is switched off as electric demand and electricity prices fall.

[5] The root mean square deviation is calculated with $\sqrt{\sum_{t=1}^{n_t} \left(q_{1,t} - \hat{q}_{1,t}\right)^2 / n_t}$.

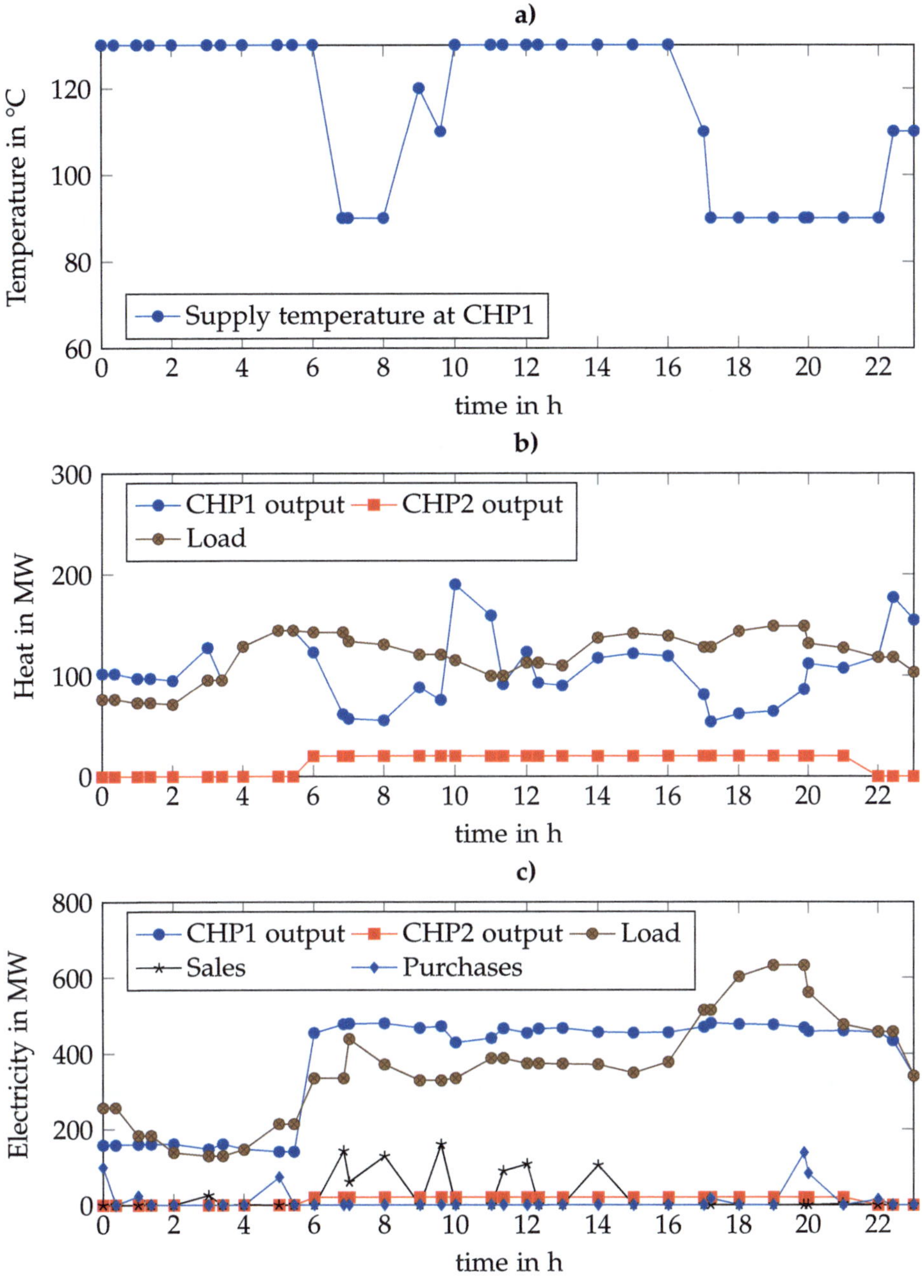

Figure 6.12: Results with exact timing for 2 CHPs scenario with hybrid discrete-continuous time model. a) Supply temperature at CHP1. b) Thermal output of CHPs in comparison to thermal load. c) Electric output of CHPs, electric load and interactions with electricity market.

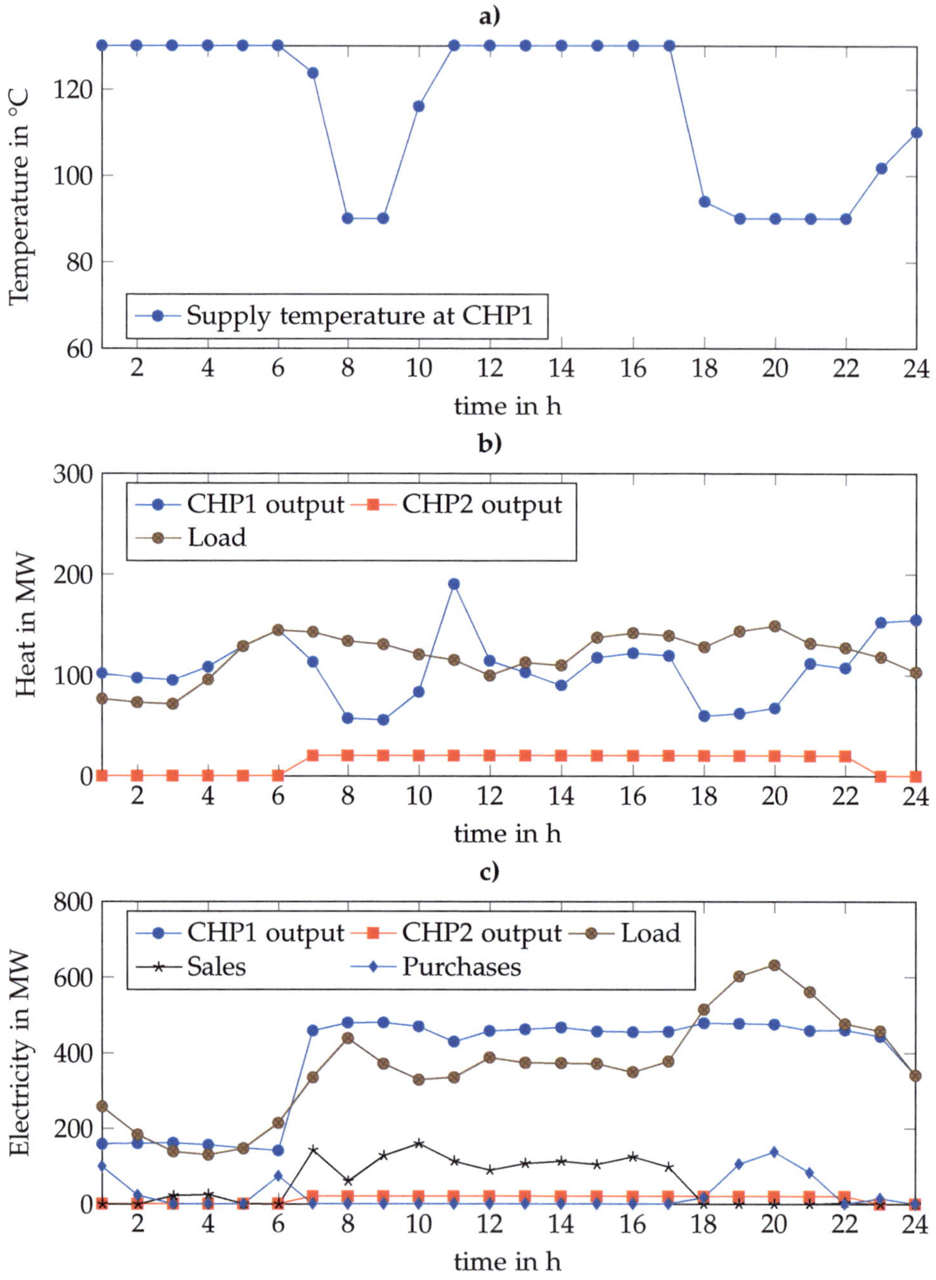

Figure 6.13: Hourly averaged results for 2 CHPs scenario with hybrid discrete-continuous time model. a) Supply temperature at CHP1. b) Thermal output of CHPs in comparison to thermal load. c) Electric output of CHPs, electric load and interactions with electricity market.

In comparison to a simulation using the node method (Section 2.1.2) with the optimization results of supply temperature of CHP1, thermal output of CHP2 and thermal load as input, the thermal output of CHP1 in the optimization result has a root mean square deviation of 32.850.

6.1.4 Optimization results with delay matrix approach

In the following, the case study results of an optimization of the heating grids described in Section 6.1.1 with the delay matrix approach introduced in Chapter 5 are presented. As in the previous section, thermal losses are not considered being of minor importance for the studied cases.

The computations for the delay matrix approach were run on the setup used in the previous section (an Intel Silver 4110 CPU with 2.10 GHz and 16 GB RAM using Julia/JuMP [Ran16] for modeling and Gurobi [Gur16] as MILP solver).

One CHP case

Figure 6.14 shows the results of the delay matrix approach for the one CHP case being achieved in 0.035 s leading to an overall cost of 291,173.4 €.

The electric output of the CHP presented in Figure 6.14 c) alternates between high and low electric output in the early morning hours until hour 6. As for the previous optimization results this is caused by an electricity market price swinging around the marginal production cost. In Figure 6.14 b) there is a thermal discharge (CHP output below load) of the grid around hours 4, 8, 12 and 18, being the high price hours. The grid is charged (CHP output above load) around hours 2, 6, 11 and 13, where electricity prices are lower. This charging and discharging is reasonable as a lower thermal output allows the CHP to increase electric output (see Figure 2.1) and, thus, revenues are increased as electric production is moved to higher price times. This load shifting can as well be observed in Figure 6.14 a) showing the supply temperature at the CHP. The supply temperature is increased to charge the grid and decreased to discharge the grid. At the end of the time horizon after the last price peaks, the supply temperature remains at the lowest level, discharging the thermal energy stored in the heating grid completely. This is reasonable as the model formulation does not account for thermal energy stored in the heating grid at the end of the time horizon.

A root mean square deviation of 49.159 for the thermal output of the CHP is obtained when comparing the optimization result to a simulation that uses the node method (Section 2.1.2). The inputs of this simulation are the optimization result of the supply temperature of the CHP and the thermal load profile.

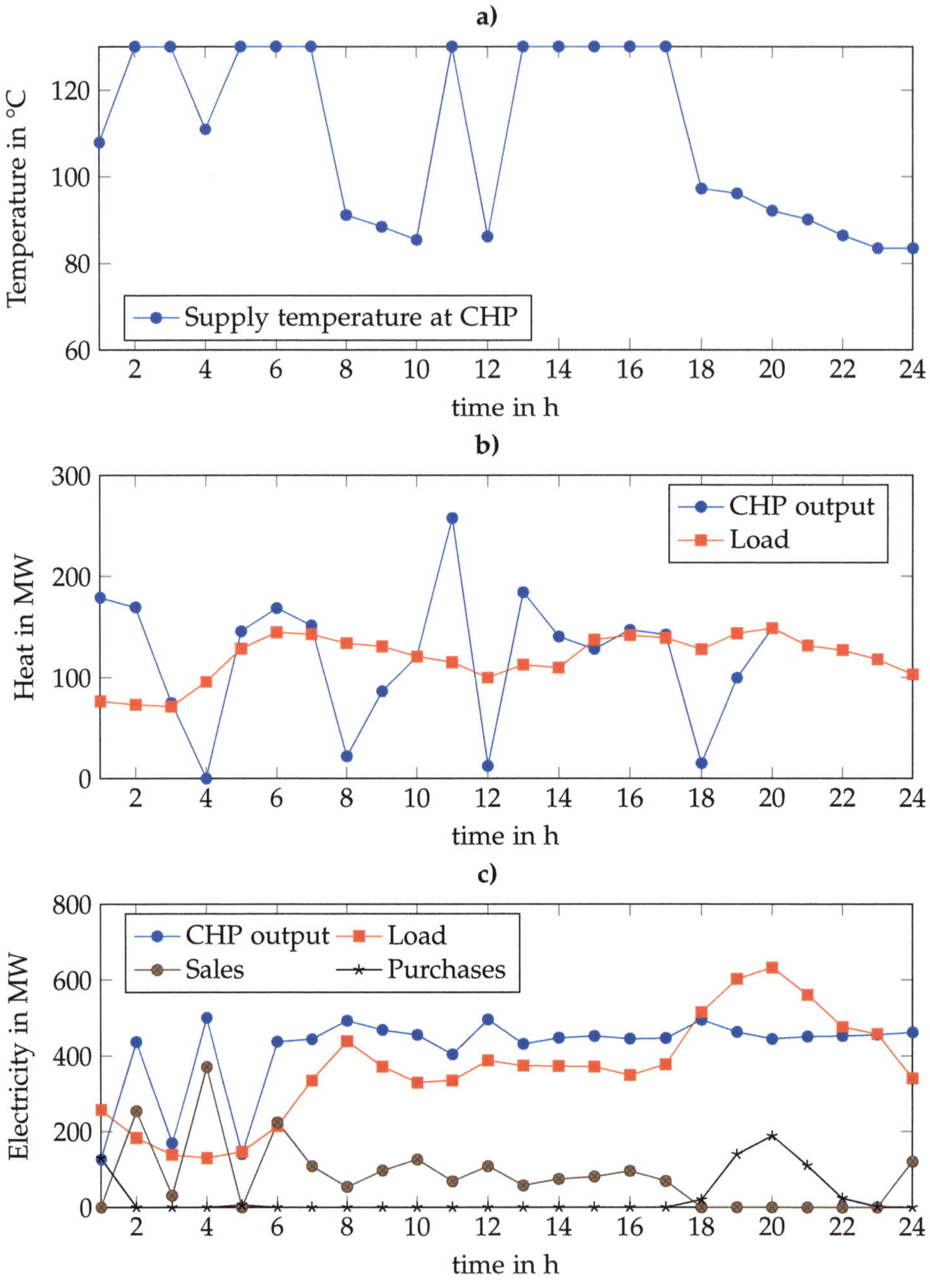

Figure 6.14: Results for 1 CHP scenario with delay matrix approach. a) Supply temperature at CHP. b) Thermal output of CHP in comparison to thermal load. c) Electric output of CHP, electric load and interactions with electricity market.

Two CHPs case

The optimization result for the two CHPs scenario as introduced in Section 6.1.1 using the delay matrix approach (Section 5) is shown in Figure 6.15. This result with an objective value of 349,537 € was achieved in 111 s having a MILP gap of 0.0044 %.

Comparing Figure 6.15 with the results for the one CHP scenario in Figure 6.14 we observe a similar pattern of the supply temperature: Before hours 8, 12 and 18 the supply temperature of CHP1 is increased to store thermal energy in the heating grid, whereas at times 8 to 10, 12 and starting in hour 18 the supply temperature of CHP1 drops back to its original level, discharging the heating grid. In these periods of discharging thermal energy from the heating grid the thermal output of CHP1 can be reduced and its electric output can thus be increased (see Figure 2.1). The overall benefit is increased by this charging pattern increasing the electricity output at higher price times. For CHP2 there is a linear connection between electric and heat output. Thus, it runs at all times when it is profitable to sell electricity to the market, except in hour 11 where as much energy as possible should be stored in the heating grid by CHP1. In hour 7 the electricity price jumps above the cost of production of CHP1 and, thus, it increases electric output to its maximum level.

Running a simulation of the heating grid case using the node method (Section 2.1.2) with optimization results of supply temperature of CHP1, thermal output of CHP2 and thermal load as input, results in a root mean square deviation of 51.160 for the thermal generation of CHP1 between simulation and optimization result.

6.1.5 Reference case: sequential approach

As a reference case we use the approach proposed in [SLM+17] to compare its performance with the approaches of this thesis. In this work, the optimization problem is split into, first, a MILP unit commitment problem that does not consider thermal dynamics of the heating grid and, second, a NLP economic dispatch problem with fixed unit commitment but considering the thermal dynamics of the grid [SLM+17]. For the unit commitment problem the formulation presented in Section 2.1.1 is used. As our cost function is linear without quadratic terms we get a MILP formulation of the unit commitment problem in comparison to [SLM+17] using a quadratic cost function. For the economic dispatch problem with fixed unit commitment the primal problems of the global optimization presented in Section 3.2 are used. Like for the global optimization, thermal losses are considered in the NLPs.

The reference case is run on an Intel Xeon CPU E5-2650 v3 with 2.3 GHz, 16 cores and 16 GB RAM using the modeling environment Julia/JuMP [Ran16] with Gurobi 8.1 [Gur16] as solver for the MILPs and IPOPT [WB06] as local solver for the NLPs. This setup was used for the global optimization case study, too. The results for the sequential approach are achieved in at most 30 min. Major solution time is required

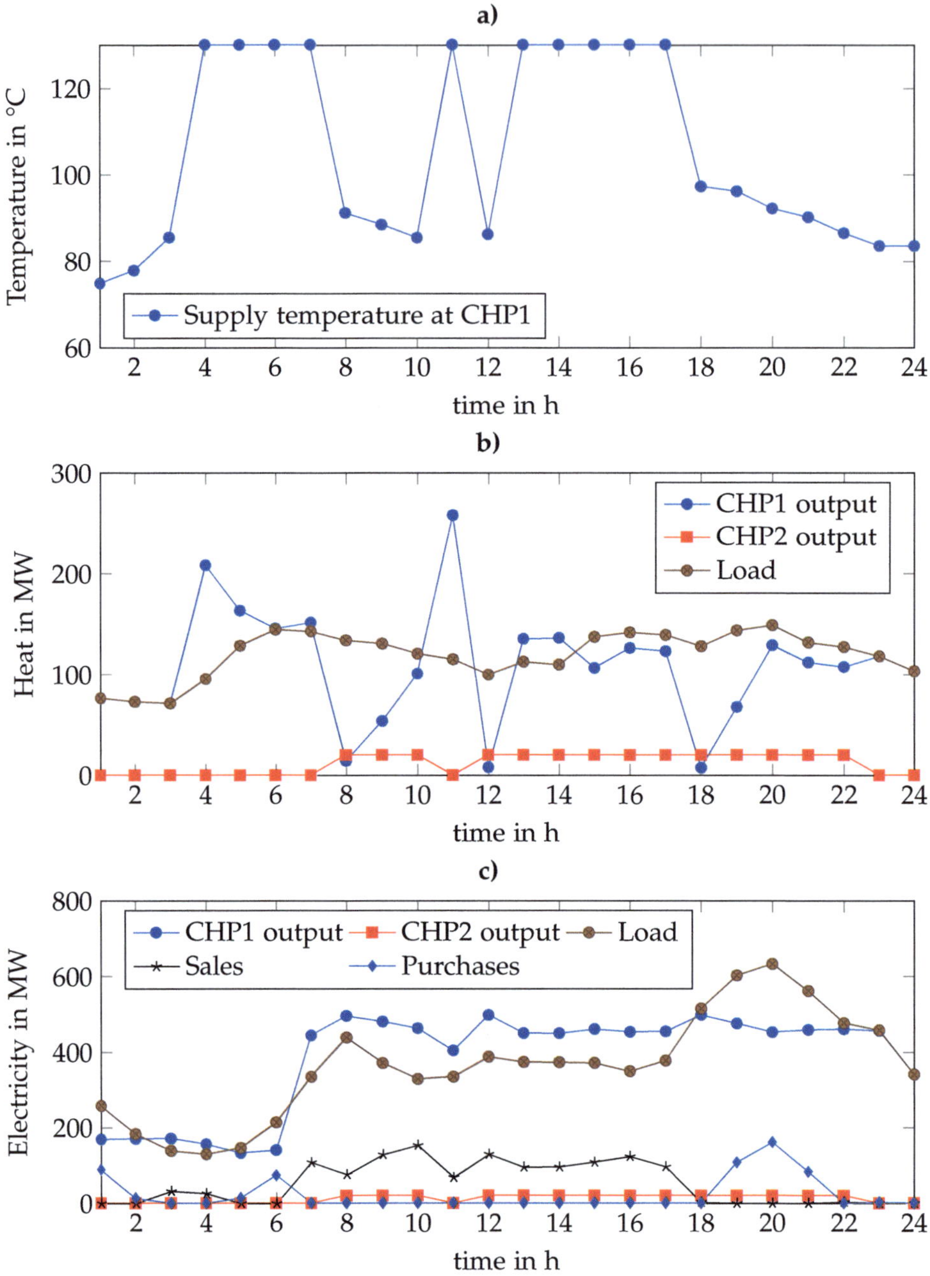

Figure 6.15: Results for 2 CHPs scenario with delay matrix approach. a) Supply temperature at CHP1. b) Thermal output of CHPs in comparison to thermal load. c) Electric output of CHPs, electric load and interactions with electricity market.

for the NLPs. This is significantly longer than the NLP solver runs in the global optimization scheme due to worse initial values. In the global optimization scheme, the initial values are taken from the results of the dual solution and, hence, are closer to the solutions than with standard initial values. This case study for the reference case is published in [MLH19].

One CHP case

If only one CHP is available, it needs to run throughout the full optimization horizon as the CHP is the only source to supply the heat demand. Thus, the CHP is always running and solving a unit commitment optimization is obsolete. The economic dispatch problem is run directly without any special initial conditions for IPOPT [WB06]. The results for the optimization runs for the finite-volumes model (Model 3.2) as well as the hybrid pipe model (Section 3.3) approach the same solutions as the corresponding global optimization scheme using those models as primal problems.

Accordingly, Figure 6.6 and Figure 6.7 showing the result of the global optimization, display the results of the reference case using the corresponding primal problem as economic dispatch problem, too. Running only the economic dispatch NLP takes about 2 to 3 seconds.

Two CHPs case

For the scenario with two CHPs the sequential approach following [SLM$^+$17] does not yield a feasible solution. The MILP unit commitment problem solves without any issues, but when running the economic dispatch NLPs, the local solver (IPOPT [WB06]) always converges to an infeasible solution for both NLP formulations, the finite-volumes model (Model 3.2) as well as the hybrid pipe model (Model 3.3). This is most likely caused by the start values of IPOPT: As the unit commitment problem does not compute any suitable starting values for the economic dispatch problem besides the commitment status, standard start values are taken. In contrast, the primal problems in the global optimization are initialized with the results of the dual problem. Hence, this issue does not arise in the cases presented in Section 6.1.2 which use multiparametric delay modeling with the same non-linear model formulations.

This situation might improve with an algorithm finding better start values. However, the result will differ from the results of the global optimization, as the commitment status of CHP2 after the unit commitment optimization in the sequential approach is different from the commitment status of CHP2 achieved in the global optimization runs as shown in Figure 6.16.

To further compare the sequential approach with the global optimization algorithm an additional case study with both CHPs at the location of CHP1 is run. The other parameters of the two CHP scenario remain. Thus, CHP1 and CHP2 are jointly feeding

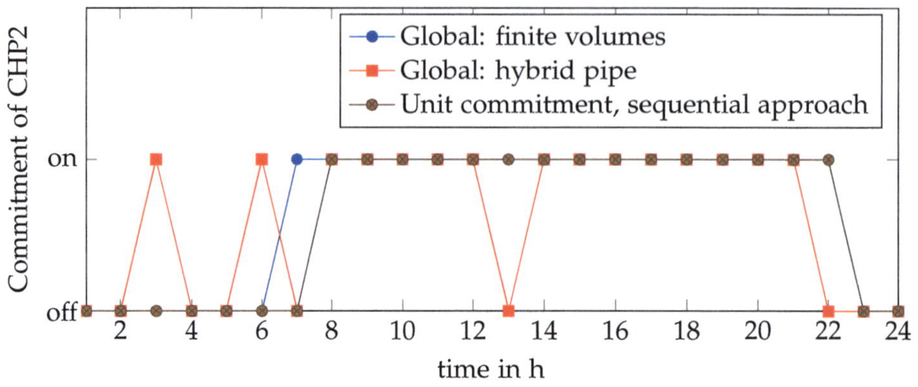

Figure 6.16: Comparison of unit commitment of CHP2 between reference case and global optimization approaches

into the 10 km pipe, which transports the thermal energy to the load area. Here, for the global optimization with both primal models, we get the same unit commitment decision as with the unit commitment in the sequential approach. In addition, with the finite-volumes model in the sequential approach the economic dispatch converges to the same overall solution as the global optimization scheme with the finite-volumes model as primal problem. However, when using the hybrid pipe model for economic dispatch in the sequential approach and as primal problem in the global optimization scheme, the algorithms result in different solutions. As the NLPs are only solved with a local solver, this is caused by the different starting values for the optimization of the hybrid pipe model in the sequential scheme and in the global optimization scheme.

6.1.6 Discussion of results and comparison of modeling approaches

Overall, the optimization results of the different approaches show several similarities: Electric generation of the CHP turns profitable in hour 6 in the one CHP case such that its electric output is increased to the maximum possible value for the current thermal output. In the two CHPs case the electricity generation becomes profitable one hour later in hour 7. In the one CHP scenarios there is a large fluctuation of the electric generation in the early morning hours as the electricity price is swinging around the marginal cost independent of the optimization approach.

Furthermore, in all results, the thermal output of the coal fired power plant is reduced around high price hours (hour 8 and hour 18) to allow a higher output of electric energy. Here, the thermal load is supplied by thermal energy stored in the heating grid by an increase of supply temperature in previous time steps. The supply temperature is decreased to its minimum level in order to discharge the thermal energy stored in the heating grid. The gas fired power plant in contrast is run in all higher price hours at maximum capacity as electric and heat output are directly linked.

With all methods, the optimizer pushes the supply temperature at the CHP to its boundaries leading to large supply temperature changes. In real-world situations large variations in temperature are usually not wanted as larger variations in temperature cause higher changes in volume requiring bigger buffer tanks and can increase material fatigue.

Table 6.2: Results overview for one CHP case

	Delay matrix	Hybrid time grid	Global: finite-volumes	Global: hybrid pipe
Temperatures in °C	60 to 130 (Continuous)	90, 95, 100, 105, 110, 115, 120, 130	60 to 130 (Continuous)	60 to 130 (Continuous)
Objective in €	291,173.4	292,565.9	291,235.7	294,386.0
Lower Bound in € (global)	-	-	288,371.3	288,265.1
Solution Time in s	0.035	800	146.19	125.00
MILP Gap in %	0	0.0513	-	-
Global Gap in %	-	-	1.0	2.1
Number of Variables (MILP)	403 Cont. 124 Int.	3,481 Cont. 1,319 Int.	8,738 Cont. 11,026 Int.	19,764 Cont. 11,026 Int.

Table 6.3: Results overview for two CHPs case

	Delay matrix	Hybrid time grid	Global: finite-volumes	Global: hybrid pipe
Temperatures in °C	60 to 130 (Continuous)	90, 95, 100, 105, 110, 115, 120, 130	60 to 130 (Continuous)	60 to 130 (Continuous)
Objective in €	349,537	346,490	348,252.9	351,112.0
Lower Bound in € (global)	-	-	330,871.1	334,785.9
Solution Time in s	111	800	1,637.0	2,907.1
MILP Gap in %	0.0044	0.0554	-	-
Global Gap in %	-	-	4.99	4.6
Number of Variables (MILP)	506 Cont. 447 Int.	3,581 Cont. 1,462 Int.	9,489 Cont. 11,676 Int.	11,337 Cont. 15,076 Int.

Comparing the results of the different methods for both cases studied in Table 6.2 and Table 6.3, we can observe that the objective values have the same order of magnitude and vary comparably little. This shows, that thermal losses, considered in the global optimization approaches, but not considered in the delay matrix and the hybrid time grid approach, are of minor importance for the studied scenarios with one pipe with a large diameter. The lower bounds achieved with the global optimization approaches are comparable, too. For the one CHP case, the global optimization approaches show very good performance. For the two CHPs case, solving the dual problem becomes a lot more challenging such that solution time and optimality gap increase. Both however stay at a reasonable level.

Assuming delay times are given, the delay matrix approach results in the smallest optimization model offering the fastest solution times. However, the extension to a second CHP presented in Section 5.3.2 increases the solution time by multiple orders of magnitude as it introduces many binary variables. Nevertheless, it still remains the fastest solver for the two CHPs case. Problem size as well as solution times for the hybrid time grid model are larger but still within an acceptable range for day ahead planning problems. As this approach offers an exact modeling of the transport delays, it produces results closer to the real-world situation as the delay matrix approach. However, possible temperatures are limited to a discrete set of temperature levels. For the case studies, steps of 5 K are assumed which should be accurate enough for most real-world applications. This was validated with a simulation model: To evaluate solution accuracy the optimization result of the thermal output of CHP1 was compared with a detailed simulation using the optimized supply temperature profile as input. The root mean square deviation between optimization result and detailed simulation reduces from 49.159 to 29.431 for the one CHP case and from 51.160 to 32.850 for the two CHPs case, if the hybrid time grid model is used instead of the delay matrix approach.

The temperature dynamics in the primal solutions using the finite-volumes model in the global optimization scheme are not completely realistic, as the outflow temperature of the pipe varies less than the inflow temperature due to the smoothing effects of the chosen models. The hybrid pipe model leads to a more accurate representation of the transport delay and to a larger objective value as shown in Table 6.2 and Table 6.3. This is reasonable as a higher variability of the CHP supply temperature allows to reach a better objective value of the primal problems with the finite-volumes model.

The sizes of the dual problems in the global optimization algorithm strongly depend on the number of iterations the algorithm needs to achieve a good solution. Solution times for the one CHP case are competitive in comparison to the hybrid time grid approach. The solution gap varies from 1.0 % to 2.1 % depending on the chosen primal problem being achieved in less than 2.5 min. However, for the two CHPs case solution times increase significantly even though the accepted global optimality gap is increased to 5 %. About 30 min and 50 min are needed to reach this gap with the different primal problems. If the MILP solver time limit for solving the lower bound problem is increased, the global optimality gap can be closed further. Nevertheless, starting from iteration 3 the MILP solver struggles to find a feasible solution for the lower bound problem. Thus, warm-starting the solving of the lower bound problem might improve this situation. Especially when larger heating grids shall be optimized, such improvements to the algorithm are required.

A unit commitment optimization using the basic MILP model presented in Section 2.1.1 not considering neither thermal dynamics of the heating grid nor thermal losses reaches an objective value of 297,788.6 € for the one CHP case and an objective value of 355,059.0 € for the two CHPs case. Thus, heating grid operators could achieve cost savings between about 3,500 € and 6,700 € per day translating to relative savings of about 1.2 % to 3.3 % using the heating grid as thermal storage.

6.2 Real-world case study

One of the goals of this thesis is to find optimization approaches for district heating grids which can directly be applied to real-world cases. As the delay matrix approach introduced in Chapter 5 enables a fast and efficient optimization of large grids, it is applied to the district heating grid of the city of Kiel, Germany in this section. This real-world case study is published in [MKV+19].

All calculations of this real-world case study are run on an Intel Xeon v2 with 2.6 GHz (2 CPUs) and 4 GB RAM using the modeling environment Julia/JuMP [Ran16] and MILP solver Gurobi [Gur16].

6.2.1 The district heating grid of Kiel

Kiel is a city at the German coast of the Baltic Sea with about 250,000 inhabitants. The district heating grid of Kiel has a total pipe length of about 370 km serving about 7,000 building connection points [Sta15b]. With this size it is a very representative large heating grid (defined as grids above 100 km length) being only slightly longer than the average large heating grid in Germany [SDEW12]. The largest heat supplier in Kiel was a coal fired power plant with about 354 MW electric output which was replaced by a gas engine plant with twenty gas combustion engines each having an electric and heat output of about 10 MW in 2019 [Sta15b, Sta16].

For this case study, 37 days in the periods January to March and November to December are selected from a full year of measurement data as in [MKV+19]. These days have been chosen, as here the coal fired power plant supplies almost all heat demand in the heating grid and, thus, flow directions are comparably stable. Therefore, it is possible to identify a stable delay matrix. Based on measurement availability, several consumption zones are introduced which split the demand of the heating grid among them. For each consumption zone, its share of the overall consumption as well as the transport delay from the generation site are calculated allowing to create the delay matrix $M[l,t]$ introduced in Table 5.1 of Chapter 5. The transport time from generation site to the main consumption area including the city center is about one to two hours for the studied days. Usually, all consumption areas in the heating grid are reached by the hot water after at most 8 hours.

The formulation introduced in Section 5.3.1 is used to consider the additional heat losses caused by the supply temperature increases chosen in the optimization. The heat loss factor α_{loss} is calculated using topology information giving installed lengths and diameters of all pipes in the heating grid in combination with data sheet information on the heat loss coefficient of the pipes. For some (mostly older) pipes, which have been insulated during the installation works, no data sheet information is available. Thus, the heat loss coefficient largely varies and is worse than the heat loss coefficient of pre-insulated pipes. Based on consultations with Stadtwerke Kiel, the heat loss coefficient of those pipes is assumed to be the double of the heat loss coefficient of a pre-insulated pipe with the same diameter.

For every day considered, prices from the German/Austrian price zone of the EPEX spot day ahead market [EPE] are taken as electricity prices for sales and with a minor offset for purchases. In the optimization these hourly varying prices are considered as parameters.

Two different generation scenarios are used, which are inspired by past and new main generators in the heating grid of Kiel, but do not reflect the exact real-world size and setting. As in both scenarios only one CHP supplies the full demand, the thermal output of the original CHPs is increased by adjusting size or number of units. In the first scenario, the district heating grid is supplied by a coal fired power plant with an extraction condensing turbine, 130 °C maximum supply temperature and maximum output of 300 MW electric and 300 MW thermal power. In part-load operation, only a minor drop of efficiency is assumed. As an extraction condensing turbine is installed, electric and thermal output are linked with the feasible operation region shown in Figure 2.1.

In the second generation scenario the district heating grid is supplied by a gas engine plant consisting of thirty gas combustion engines each having a maximum thermal output of 10 MW and a maximum electric output of 10 MW. The gas combustion engines supply heat at a maximum temperature of 115 °C having a linear power-to-heat ratio. In part-load operation, a linear efficiency drop is assumed, having the highest efficiency at maximum output. The start up cost of the gas engines is assumed to be low. An overview on the different parameters of the generation scenarios used in the real-world case study is given in Table 6.4.

Table 6.4: Generation scenarios in real-world case study

	Coal fired plant	Gas engine plant
Number of units	1	30
Electric output per unit in MW	≤ 300	≤ 10
Thermal output per unit in MW	≤ 300	≤ 10
Supply temperature in °C	≤ 130	≤ 115
Part-load efficiency drop	small	big
Power to heat ratio	see Figure 2.1	constant

6.2.2 Results for a December day

To show detailed results for one day, the delay matrix approach introduced in Section 5 is run for a December day for both generation scenarios presented in the previous section. Using measurement data of the district heating grid of Kiel from this date, the overall thermal energy consumption profile as well as the share of consumption of the different consumption zones and the time delay from thermal generation to these consumption zones are calculated. This information is used to parameterize the delay matrix $M[l,t]$. All electric energy produced by the CHPs is sold at the energy market. Figure 6.17 shows the relative price profile for this December day scaled with the price peak of the day.

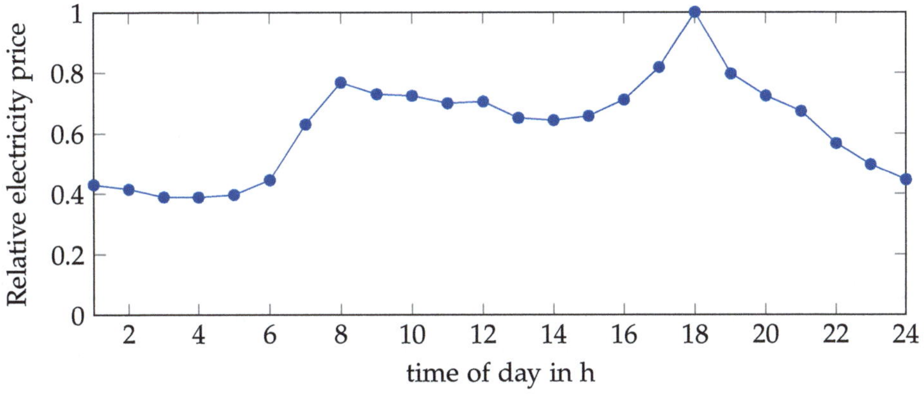

Figure 6.17: Relative electricity price for a December day

It is assumed that the supply temperature before optimization is at the minimum possible temperature according to the technical connection conditions of the heating grid. Thus, the supply temperature profile before optimization is calculated based on the outdoor air temperature. For the second generation scenario with the gas engine plant it is additionally assumed that a temperature increase cannot exceed 10 K.

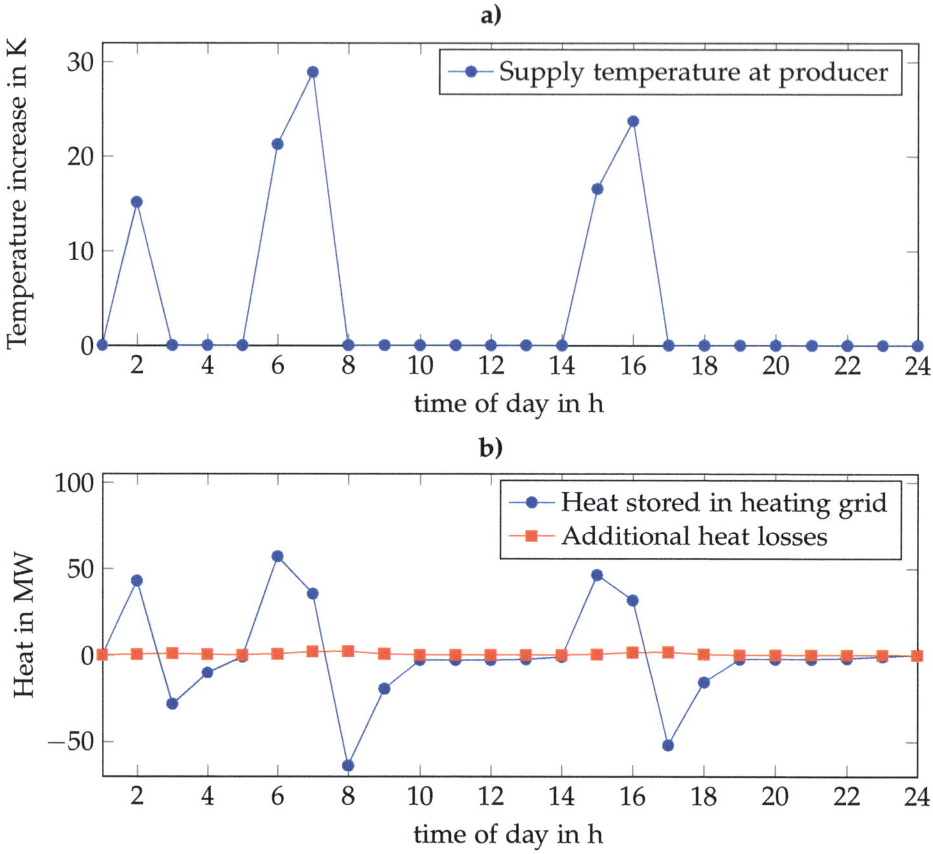

Figure 6.18: Result of real-word case study for coal fired power plant (one day). a) Supply temperature increases. b) Thermal energy stored in the district heating grid with supply temperature variations and additional thermal losses caused by supply temperature increase.

The optimization result for the first generation scenario consisting of the coal fired power plant with extraction condensing turbine is shown in Figure 6.18. Thermal energy is stored in the heating grid at hour 2, hours 6 to 7 and hours 15 to 16 with an increase of supply temperature at the producer. These periods are followed by a discharge of thermal energy from the heating grid. During this discharging phase, the thermal output of the coal fired CHP is reduced enabling a higher output of electricity of the CHP according to the feasible operation region of this CHP (Figure 2.1). Thus, the supply temperature increases to charge the heating grid before peaks of electricity price around hours 8 and 18. During these peak times the thermal energy stored within the heating grid is used to supply the thermal demand and the CHP can increase its electric output. Hence, electric generation of the CHP is moved to hours where higher revenues are possible. The increase of thermal losses is comparably low.

Nevertheless, it reduces overall profits of storing thermal energy in the heating grid. This optimization using the delay matrix model finished in less than 5 ms increasing the overall profit of operations by several hundred Euros (~0.2 %) in comparison to a scenario without thermal storage.

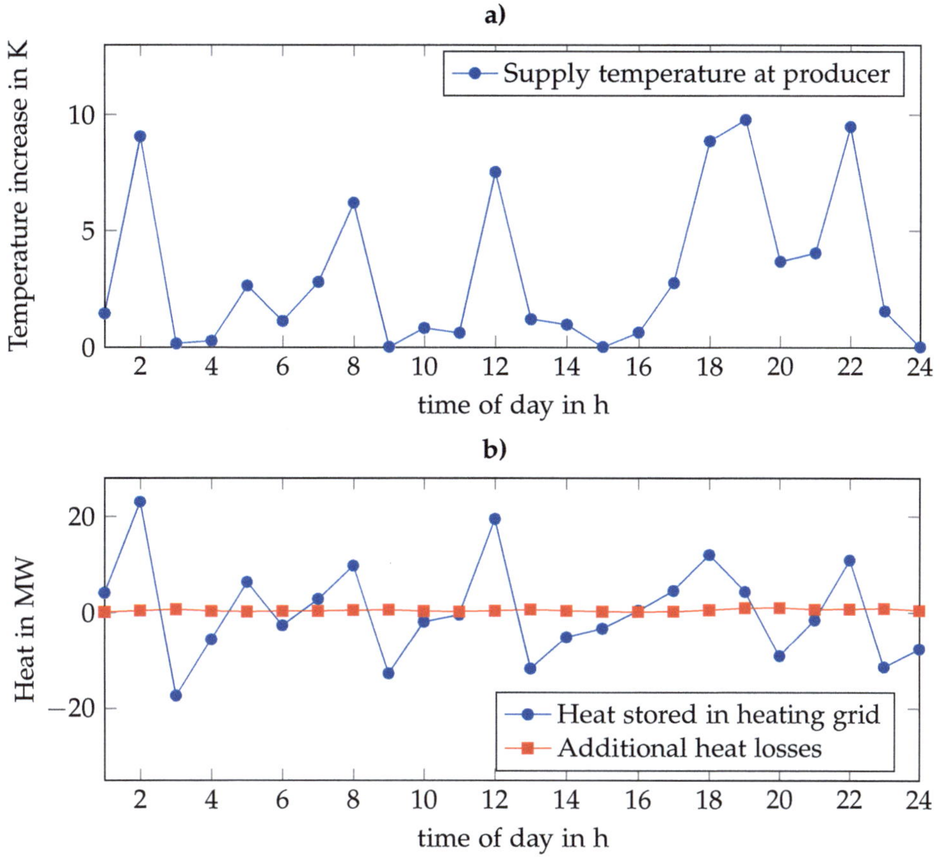

Figure 6.19: Result of real-word case study for gas engine plant (one day). a) Supply temperature increases. b) Thermal energy stored in the district heating grid with supply temperature variations and additional thermal losses caused by supply temperature increase.

The result for the second generation scenario consisting of a gas engine plant with 30 individual gas combustion engines is shown in Figure 6.19. In contrast to the coal fired plant, output of thermal energy increases with output of electricity for the gas engine plant. Hence, the optimizer increases supply temperature at high price times around hours 8 and 18 storing thermal energy in the grid and increasing power output of the CHP. However, there are many other variations in supply temperature which cannot be explained with the volatile energy price. They are caused by the drop of efficiency in part-load operations. As it is more efficient to run a gas combustion engine at maximum speed, the optimizer pushes to run gas units at maximum output if they are running. But as the thermal demand is usually not at an integer multiplier of the gas engine size, the thermal storage capacity of the heating grid is used to buffer the differences. Thermal losses are less than for the coal fired power plant case as the temperature increases are smaller. Despite these rather small supply temperature increases, profits for the gas engine plant case are higher than for the coal fired power plant case, achieving savings of several thousand Euros in this day (~2.5 % of overall generation cost). Looking at the end of the optimization horizon it must be noted that there is a rather late big supply temperature increase in hour 22, such that the heating grid is not fully discharged at the end of the optimization horizon. Thus, if this thermal energy could be used e.g. extending the optimization horizon, savings might increase further. As more integer variables are involved, the optimization of the gas engine plant case takes 1.7 s which is several orders of magnitude longer than the optimization of the coal fired power plant case. Nevertheless, this remains an acceptable solution time.

6.2.3 Evaluation of economic potential

To evaluate the economic potential of using the heating grid as thermal storage, the 37 days with feasible measurement data of the heating grid of the city of Kiel are grouped into several scenarios. Grouping by different optimization horizons we get

- 37 scenarios for a one day (24 h) horizon,
- 27 scenarios for a two day (48 h) horizon,
- 19 scenarios for a three day (72 h) horizon and
- 14 scenarios for a four day (96 h) horizon.

Note that the scenarios are partly overlapping for multi day scenarios. Several cases are studied running optimizations for both generation setups presented in Section 6.2.1 and for all scenarios mentioned above.

For the coal fired power plant the following four cases are run with differences in the supply temperature level assumed before optimization and in consideration of thermal losses.

- minimum supply temperature without thermal losses
- minimum supply temperature with thermal losses
- measured supply temperature without thermal losses
- measured supply temperature with thermal losses

The minimum supply temperature is calculated with the outdoor air temperature according to the technical connection conditions of the heating grid of Kiel [Sta15a]. Measured supply temperature cases take the actual measurement of supply temperature at the CHP in the scenario as optimization input.

For the gas engine plant three cases are studied. All of them consider thermal losses and are using the minimum supply temperature calculated with the outdoor air temperature according to the technical connection conditions of the heating grid of Kiel [Sta15a]. It is not feasible to use the measured temperature profiles as they often are above the maximum supply temperature of the gas engine plant. The cases studied for the gas engine plant are:

- original part-load efficiency
- higher part-load efficiency
- original part-load efficiency and limitation of supply temperature increase to 10 K

Coal fired power plant

An overview on the case study results for the coal fired power plant is given in Figure 6.20. Comparing these results, three major influences on the relative savings of using the heating grid as thermal storage in comparison to a scenario without thermal storage can be identified.

First, it can be noted that increasing the optimization horizon increases the average and decreases the variance of the savings for all cases. With a longer time horizon the variance in the electricity prices between the scenarios decrease and, accordingly, the variance of the relative savings decreases: If there is one very favorable electricity price increase or decrease in a day, its impact is reduced, if not only this day, but also the neighboring days without such a favorable electricity price development are considered. The increase of the average and median relative savings with increasing time horizon is most likely caused by inefficient usage of the thermal storage capability of the heating grid at the end of the optimization horizon: If the heating grid is not fully discharged at the end of the optimization horizon, the remaining thermal energy is considered lost in the optimization. Thus, storing energy at the end of the optimization horizon is less interesting and changes in electricity price at the end of the optimization horizon cannot be used efficiently. Having an optimization horizon spanning multiple days allows to leverage the variations of electricity price in the evening for all but the last day without limitations. Thus, the negative impact of this problem decreases.

Second, it can be observed that the cases with minimum supply temperature are creating higher benefits than the cases with measured supply temperature. As the measured supply temperature always lies above the minimum supply temperature, the maximum increase of supply temperature before reaching the technical limitations

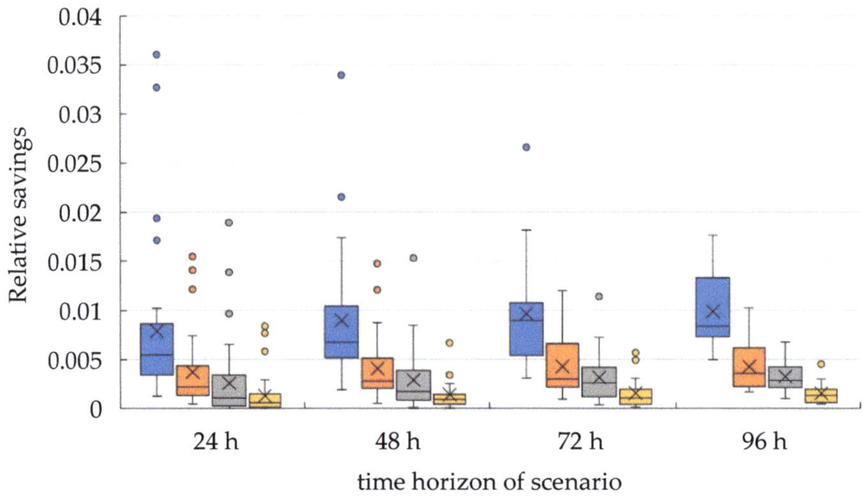

Figure 6.20: Relative savings achieved in real-world case study for the coal fired power plant for different time horizons and cases: minimum supply temperature without thermal losses (blue) and with thermal losses (gray); measured supply temperature without thermal losses (orange) and with thermal losses (yellow). The boxes mark the lower and upper quartiles, hence 50 % of the data lies inside the box. The crosses mark the average and the line inside the box the median of the data. The whiskers show the variability of data within at most a distance from lower and upper quartile of 1.5 times the inner quartile range. Points denote outliers being data outside this range.

of the coal fired power plant is smaller. The thermal energy storage capacity of the heating grid is proportional to the possible increase of supply temperature. Thus, having a higher supply temperature before the optimization, reduces the thermal storage capacity of the heating grid and, thus, less savings can be achieved.

Third, considering heat losses reduces the economic benefits. This connection is expected, as considering heat losses requires additional thermal generation increasing cost without increasing revenues. As can be seen in the one day example in Section 6.2.2, temperature increases in the coal fired power plant case can be high (up to 30 K). Thus, the increase of thermal losses caused by this temperature increase reduces the relative savings. Thermal losses induced by temperature increases play an important role for the considered heating grid in the city of Kiel, as the temperature increases reach all pipes in the system, even the very small ones connecting single houses. Hence, the surface of the heating grid network is very large in comparison to its volume. For heating grids with only larger pipes the importance of thermal losses is lower. The same applies to heating grids where the temperature in large pipes and small pipes is decoupled. Here the storing of thermal energy can be limited to the larger pipes where temperature increases lead to smaller thermal losses than in small pipes with high surface to volume ratio.

The solution times for all scenarios with the coal fired power plant are extremely fast, never exceeding 50 ms.

Gas engine plant

Figure 6.21 shows an overview on achievable relative savings based on different scenarios and cases for the gas engine plant. Only cases considering thermal losses and assuming the minimum supply temperature as pre-optimization supply temperature are calculated for the gas engine plant. A similar trend as for the coal fired generation can be observed regarding an increase of time horizon: In scenarios with a longer time horizon, the average and median achieved relative savings increase and the variance of the relative savings decrease. However, the second effect is less important, as for the gas engine plant major savings are achieved running the combustion engines at maximum efficiency rather than shifting generation based an varying energy prices as discussed for the one day scenario in Section 6.2.2.

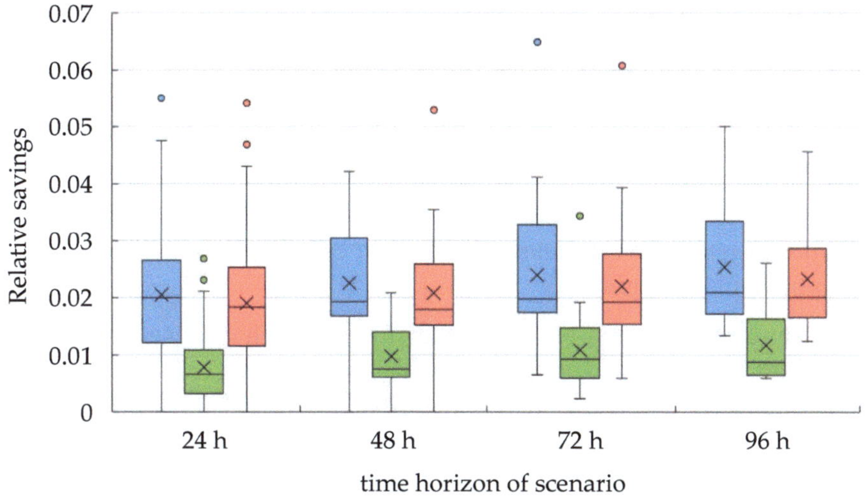

Figure 6.21: Relative savings achieved in real-world case study for gas engine plant for different time horizons and cases: original part-load efficiency (light blue); higher part-load efficiency (green); original part-load efficiency and maximum temperature increase of 10 K (red). The boxes mark the lower and upper quartiles, hence 50 % of the data lies inside the box. The crosses mark the average and the line inside the box the median of the data. The whiskers show the variability of data within at most a distance from lower and upper quartile of 1.5 times the inner quartile range. Points denote outliers being data outside this range.

This implication, that the main benefits for the gas engine plant scenarios are drawn from running the individual gas units at maximum efficiency, is supported if the cases with original part-load efficiency and higher part-load efficiency are compared. The achieved benefits for the gas engine plant with higher part-load efficiency are significantly smaller than the relative savings achieved with original part-load efficiency. Thus, for the gas engine plant cases using all running units at maximum efficiency is the major reason for the savings generated.

In the third case studied for the gas engine plant, the temperature increase was limited to 10 K. Here, the achieved relative savings are only slightly reduced. Thus, the temperature increases required to achieve benefits with the gas engine plant are comparably small and a lot less important than the temperature increases calculated for the coal fired power plant. This is supported by the results for one example day discussed in Section 6.2.2. Despite this fact, the savings generated using the heating grid as a thermal storage with the gas engine plant are importantly higher than the savings generated with the coal fired power plant shown in Figure 6.20. This is remarkable as thermal losses are considered for all gas engine scenarios. The major reason for this is most likely that the coal fired plant can generate benefits only by shifting generation between different price time slots, which requires a rather large thermal storage. The gas engine CHP can additionally generate savings by buffering thermal energy to run its units at maximum efficiency if they run, which requires less thermal storage. The gas engine plant has also a more direct link between thermal and electric output than the coal fired CHP. Hence, changes in thermal output of the assumed gas engine plant have stronger influence on its electric output than for the coal fired power plant. Thus, smaller changes in thermal generation are required to achieve the same change in electric output.

Note that part-load efficiencies and generation cost of the gas engine plant and the coal fired power plant are estimated. Real-world behavior of a gas engine plant or a coal fired power plant might vary.

Finally, in Figure 6.21 it can be noted that for the 24 h and 48 h scenarios there are always some scenarios where no benefits can be achieved at all. This can be explained by Figure 6.22 which shows that the days not achieving any benefits are days with very low ambient air temperatures. At low ambient air temperatures the heating grid must be supplied with high supply temperatures (at least 115 °C for ambient air temperatures below -5 °C according to the technical connection conditions [Sta15a]). Hence, there is no room for a temperature increase as the gas engine plant can only produce up to 115 °C and no thermal energy can be stored in the heating grid. Thus, the energy storage capabilities of the heating grid can mainly achieve interesting economic savings at average ambient air temperatures above 0 °C.

Solution times for the scenarios with the gas engine plant are slower than optimizations for the scenarios with the coal fired power plant, as more integer variables representing the commitment status of the individual gas engines are involved. For the scenarios and cases presented in this section they are between 1 s for the 24 h scenarios and reach up to 100 s being the MILP solver time limit for some of the 72 h and 96 h scenarios.

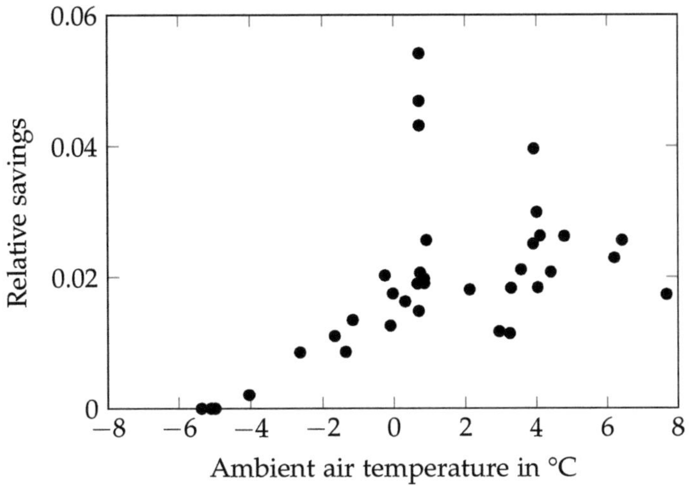

Figure 6.22: Dependency of relative savings for the gas engine plant on daily average ambient air temperatures for scenarios with 24 h horizon

6.3 Summary of chapter

This chapter contains the evaluation of the methods developed in this thesis. In Section 6.1 all optimization methods developed in this thesis were compared using two small test scenarios: one with one CHP and one with two CHPs. The delay matrix approach (Chapter 5) proves its competitiveness being the fastest optimization model in all studied cases, very often providing a solution in less than a second. Solving the hybrid time grid model (Chapter 4) takes longer, but solution times stay within an acceptable range for a day ahead planning problem being at most 10 minutes. Comparing the optimization result with a simulation using the optimization result as input, the results of the hybrid time grid model show a higher precision than the results using the delay matrix approach. This is expected as the hybrid time grid model allows an accurate representation of the time delay.

The global optimization approach (Chapter 3) was run using both introduced primal problems. Solution times of the global optimization approach are competitive only for the one CHP case. For the two CHPs case, solution times increase to 30 min and 50 min. The same applies to the global optimality gap achieved: for the one CHP case gaps of 1 % and 2.1 % are possible, whereas for the two CHPs case gaps of about 5 % are achieved. Thus, increasing the size of the studied heating grid increases the computational effort drastically. The objective value with the finite-volumes model as primal problem is always slightly smaller than the objective value with the hybrid pipe model as primal problem. This can be explained by the less accurate representation of the time delay in the finite-volumes model: As there is a smoothing of temperature introduced by the finite-volumes, the temperature at the CHP can vary more than with the hybrid pipe model (sometimes even to values not possible in real-world). Thus, variations in generation are larger allowing to create more benefits. In addition to the approaches presented in this thesis, the sequential approach proposed in [SLM+17] was run for both studied cases. This method reaches the same result as the global optimization approaches for the one CHP case, but does not succeed in delivering a solution for the two CHPs case. Overall, all approaches give comparable results, being slightly above the calculated global lower bounds.

In Section 6.2 the delay matrix approach (Chapter 5) was applied in a real-world case study using measurement data from the heating grid of the city of Kiel. The delay matrix was created using transport delays to and share of consumption of load areas, being extracted from this measurement data. A thermal loss factor was identified using grid topology and pipe data sheet information. Two different generation cases were considered: a coal fired power plant with extraction condensing turbine and a gas engine plant consisting of 30 gas engines. Different scenarios were used based on the real-world measurement data of 37 days in the periods January to March and November to December. All results were compared with an optimization not considering the storage capabilities of the heating grid and corresponding relative savings were calculated. After the presentation of results for one day in December in Section 6.2.2, a summary of all studied scenarios and cases was presented in Section 6.2.3. Several influences on the achievable savings using the heating grid as a storage have been identified:

- A longer optimization horizon allows higher savings, as inefficiencies at the end of the optimization horizon occur less often.
- Ambient air temperature has an important influence on achievable savings, because no increases in supply temperature are possible when the heating grid operates close to full load.
- Heat losses decrease the achievable savings. If smaller pipes are decoupled from large pipes (with heat exchangers), this effect can be reduced.
- Shifting load/generation between time slots requires a large amount of thermal storage, hence, temperature variations and the influence of thermal losses are big.
- Using the storage capacity to run plants at maximum efficiency requires less thermal storage. Thus, temperature variations are smaller, thermal losses are less important and savings are higher.

All in all, the delay matrix approach proved to be capable to optimize real-world heating grids. Optimization times stayed in an acceptable range mostly being solved in less than a second. Only few larger scenarios with longer optimization horizon took up to 100 s to solve. The resulting savings are up to 6 % or a few thousand Euros per day and, thus, could easily reach more than 180,000 € per winter season (assuming average savings of 1,000 € per day for half a year). However, they are dependent on the grid topology as well as on the generation setup. Hence, no general recommendation on using the heating grid as thermal energy storage can be given and an analysis of generation setup and grid topology is required to identify saving potentials.

7 Conclusions

With increasing share of renewable power generation in electric grids, conventional generation needs to become more and more flexible to react to fluctuating renewable feed in and volatile electricity prices. As heat demand in heating grids is usually not flexible, CHPs are facing major challenges. One way to enable a more flexible operation of CHPs is the installation of thermal storage tanks. A promising alternative, that avoids installation effort and investment cost, could be to directly use the ability of the thermal dynamics of the heating grid to store thermal energy. However, considering the thermal dynamics of a heating grid in operations planning of CHPs is complex due to bilinear terms and a variable-dependent time delay. Several optimization algorithms in literature address this non-convex optimization problem, but no approach reaches the global optimum. In addition, such optimization of energy flows considering heating grid dynamics is not widely used in practice.

Multiparametric delay modeling presented in Chapter 3 is the first optimization algorithm proving global optimality of its solution. It introduces a new outer approximation of the variable-dependent time delay which is formulated as MILP problem. Thus, it can be solved with off-the-shelf solvers. In combination with models of thermal dynamics from literature sources, this outer approximation is used in a primal-dual optimization scheme. Using this iterative approach, the solution converges to the global optimum. In the case studies in Chapter 6 it is shown that for a small heating grid with one CHP and one load the algorithm finds a solution with a global optimality gap of 1 % in less than two minutes. However, as soon as problem size increases (e.g. adding one CHP) solution times increase importantly. Thus, multiparametric delay modeling is a great approach to benchmark other optimization approaches with respect to global optimality, but without further improvements it is not a suitable solution for every day operations planning of a heating grid.

To reach a single, very exact model of the thermal dynamics in a heating grid, which can be used for every day operations planning, the hybrid time grid approach was presented in Chapter 4. It adapts optimization models originally developed for pipeline scheduling in the petrochemical industry. To suite the needs of heating grid optimization a hybrid discrete-continuous time grid replaces the continuous time grid of the pipeline scheduling model and discrete temperature levels for the supply temperature are introduced. This of course limits the choice of supply temperatures, but if the number of temperature levels is high enough (e.g. every 5 Kelvin) the influence of this discretization is limited. With these discrete temperature levels and the hybrid discrete-continuous time grid, a MILP model formulation of the thermal dynamics is possible. The resulting MILP optimization model for the full problem setup can be solved directly with standard solvers. Solutions times in the case studies (Section 6.1.3) are acceptable. An extension to heating grids with several interconnected pipes and reversible flow is possible, as extensions of the underlying scheduling models for pipeline scheduling already support such pipe networks. However, all pipe parameters need to be known to correctly parameterize the grid. Thus, the hybrid time grid approach seems to be very suitable for transport grids with a limited number of pipes.

For distribution grids with a high number of pipes with often unknown parameters, a different approach is needed. In comparison to most literature approaches, the delay matrix approach presented in Chapter 5 does not require a detailed model of the heating grid. It can be parameterized from measurement data and hence is ideal for brownfield installations. Only load share of and transport time to consumption areas are needed. The delay matrix approach enables very fast solution times as it assumes constant transport delays except for the extension to multiple CHPs. Here, additional binaries representing the different velocities are introduced, enabling a variable delay at the cost of slightly increased solution times. Nevertheless, compared to the other algorithms developed in this thesis it is still very fast as shown in the case study in Section 6.1. In Section 6.2 the delay matrix approach was applied to a real world heating grid. To reduce modeling efforts the delay matrix was parameterized using measurement data. The results show that the delay matrix approach can reduce operational cost by up to 6 % or several thousand Euros per day which could easily result in in savings of more than 180,000 € for one winter season. The variations of the savings are dependent on optimization horizon (the longer the better), grid topology (a few larger pipes are better than many small pipes) and ambient air temperature (at very low temperatures the heating grid is fully loaded and thermal storage is not possible). In addition, the solutions in the real world case study were almost all achieved in less than 1 s. Only few optimization runs took up to 100 s. Thus, the delay matrix approach is suitable for every day operations planning.

Taken together, this thesis hopefully enables a wider usage of scheduling of CHPs considering heating grid dynamics in heating grids worldwide to leverage its full potential of economic and environmental benefits. Besides direct application of the developed approaches, it might help researchers to improve their algorithms with a globally optimal benchmark.

References

Publicly available references

[AKF83] Faiz A. Al-Khayyal and James E. Falk. Jointly constrained biconvex programming. *Mathematics of Operations Research*, 8(2):273–286, 1983.

[AMP14] Maria Alessandra Ancona, Francesco Melino, and Antonie Peretto. An optimization procedure for district heating networks. *Energy Procedia*, 61:278–281, 2014.

[BBR95] Atli Benonysson, Benny Bøhm, and Hans F. Ravn. Operational optimization in a district heating system. *Energy Conversion and Management*, 36(5):297–314, 1995.

[BEN05] Michel Berkelaar, Kjell Eikland, and Peter Notebaert. lp solve: Open source (Mixed-Integer) Linear Programming system. `http://lpsolve.sourceforge.net/`, 2005. (accessed April 6, 2019).

[BHW17] Simon Bachler, Johannes Huber, and Frank Woittennek. Zur thermischen Modellierung inkompressibler Rohrströmungen. *at - Automatisierungstechnik*, 65(8), 2017.

[BJPS11] Daniele Basciotti, Florian Judex, Olivier Pol, and Ralf-Roman Schmidt. Sensible heat storage in district heating networks: a novel control strategy using the network as storage. In *International Renewable Energy Storage Conference*, 2011.

[Cas17] Pedro M. Castro. Optimal scheduling of multiproduct pipelines in networks with reversible flow. *Industrial & Engineering Chemistry Research*, 56(34):9638–9656, 2017.

[CC04] Diego C. Cafaro and Jaime Cerdá. Optimal scheduling of multiproduct pipeline systems using a non-discrete MILP formulation. *Computers & Chemical Engineering*, 28(10):2053–2068, 2004.

[CCMC15] Vanina G. Cafaro, Diego C. Cafaro, Carlos A. Méndez, and Jaime Cerdá. MINLP model for the detailed scheduling of refined products pipelines with flow rate dependent pumping costs. *Computers & Chemical Engineering*, 72:210–221, 2015.

[Cer02] Andreas Cerbe. *Prozessnahe Einsatzoptimierung mit BoFiT unter Berücksichtigung der Netzrestriktionen: Verbundprojekt Bessere Ausnutzung von Fernwärmeanlagen; Pilotprojekt HEW Hamburg*. Hamburg, 2002.

[CGCP16] Charles-Henri Clerget, Jean-Philippe Grimaldi, Meriam Chebre, and Nicolas Petit. Optimization of dynamical systems with time-varying or input-varying delays. In *2016 IEEE 55th Conference on Decision and Control (CDC)*, pages 2282–2289. IEEE, 2016.

[CGZ18] Pedro M. Castro, Ignacio E. Grossmann, and Qi Zhang. Expanding scope and computational challenges in process scheduling. *Computers & Chemical Engineering*, 114:14–42, 2018.

[CHG09] Pedro M. Castro, Iiro Harjunkoski, and Ignacio E. Grossmann. New continuous-time scheduling formulation for continuous plants under variable electricity cost. *Industrial & Engineering Chemistry Research*, 48(14):6701–6714, 2009.

[CPB17] Charles-Henri Clerget, Nicolas Petit, and Lorenz T. Biegler. Dynamic optimization of a system with input-dependant time delays. In *Proceedings of the Foundations of Computer Aided Process Operations / Chemical Process Control FOCAPO/CPC*, 2017.

[CT13] Pedro M. Castro and João P. Teles. Comparison of global optimization algorithms for the design of water-using networks. *Computers & Chemical Engineering*, 52:249–261, 2013.

[dACZ+17] Leonardo Salsano de Assis, Eduardo Camponogara, Bernardo Zimberg, Enrique Ferreira, and Ignacio E. Grossmann. A piecewise McCormick relaxation-based strategy for scheduling operations in a crude oil terminal. *Computers & Chemical Engineering*, 106:309–321, 2017.

[DCM+19] Yuanhang Dai, Lei Chen, Yong Min, Qun Chen, Junhong Hao, Kang Hu, and Fei Xu. Dispatch model for CHP with pipeline and building thermal energy storage considering heat transfer process. *IEEE Transactions on Sustainable Energy*, 10(1):192–203, 2019.

[DMOK19] Abdulrahman Dahash, Sebastian Mieck, Fabian Ochs, and Hans Joachim Krautz. A comparative study of two simulation tools for the technical feasibility in terms of modeling district heating systems: An optimization case study. *Simulation Modelling Practice and Theory*, 91:48–68, 2019.

[Dot97] Erik Dotzauer. *Algorithms for Short-Term Production-Planning of Cogeneration Plants*. PhD thesis, Linköping University, Linköping, Sweden, 1997.

[EPE] EPEX SPOT SE. `http://www.epexspot.com`. (accessed April 6, 2019).

[FKBV13] Rüdiger Franke, Stephan Kautsch, Marcel Blaumann, and Lothar Vogelbacher. Intraday-Optimierung, Pooling und vorausschauende Zielsollwertführung: Integration von erneuerbaren und konventionellen Energien. *BWK*, 65(9), 2013.

[FRV+18] John Forrest, Ted Ralphs, Stefan Vigerske, Lou Hafer, Bjarni Kristjansson, jpfasano, Edwin Straver, Miles Lubin, Haroldo Gambini Santos, rlougee, and Matthew Saltzman. Coin-or/cbc: Version 2.9.9. `https://projects.coin-or.org/Cbc`, 2018. (accessed April 6, 2019).

[FV93] Christodoulas A. Floudas and Vishy Visweswaran. Primal-relaxed dual global optimization approach. *Journal of Optimization Theory and Applications*, 78(2):187–225, 1993.

[GA07] Martin Geidl and Göran Andersson. Optimal power flow of multiple energy carriers. *IEEE Transactions on Power Systems*, 22(1):145–155, 2007.

[GBS+12] Ignacio E. Grossmann, Lorenz T. Biegler, Jeffrey J. Siirola, Nikolaos V. Sahinidis, and B. Erik Ydstie. CAPD short course: Optimization modeling, conceptual design and integrated process operations, 2012.

[Gir16] Loïc Giraud. *Modélisation dynamique et gestion avancée de réseaux de chaleur*. PhD, Université Grenoble Alpes, Grenoble, 2016.

[GMBV17] Loïc Giraud, Massinissa Merabet, Roland Baviere, and Mathieu Vallée. Optimal control of district heating systems using dynamic simulation and mixed integer linear programming. Linköping Electronic Conference Proceedings, pages 141–150. Linköping University Electronic Press, 2017.

[Gro12] Sebastian Groß. *Untersuchung der Speicherfähigkeit von Fernwärmenetzen und deren Auswirkungen auf die Einsatzplanung von Wärmeerzeugern*. PhD, TU Dresden, Dresden, 2012.

[Gur16] Inc. Gurobi Optimization. Gurobi optimizer reference manual. `http://www.gurobi.com`, 2016. (accessed April 6, 2019).

[GWL+17] Wei Gu, Jun Wang, Shuai Lu, Zhao Luo, and Chenyu Wu. Optimal operation for integrated energy system considering thermal inertia of district heating network and buildings. *Applied Energy*, 199:234–246, 2017.

[HH13] Hubert Hadera and Iiro Harjunkoski. Continuous-time batch scheduling approach for optimizing electricity consumption cost. In Andrzej Kraslawski and Ilkka Turunen, editors, *23rd European Symposium on Computer Aided Process Engineering*, volume 32 of *Computer Aided Chemical Engineering*, pages 403–408. Elsevier Science, San Diego, CA, USA, 2013.

[HK10] M. M. Faruque Hasan and Iftekhar A. Karimi. Piecewise linear relaxation of bilinear programs using bivariate partitioning. *AIChE Journal*, 56(7):1880–1893, 2010.

[HMB+14] Iiro Harjunkoski, Christos T. Maravelias, Peter Bongers, Pedro M. Castro, Sebastian Engell, Ignacio E. Grossmann, John Hooker, Carlos Méndez, Guido Sand, and John Wassick. Scope for industrial applications of production scheduling models and solution methods. *Computers & Chemical Engineering*, 62:161–193, 2014.

[HT96] Reiner Horst and Hoang Tuy. *Global Optimization*. Springer Berlin Heidelberg, Berlin, Heidelberg, 1996.

[KC13] Manfred Klöpsch and Jürgen Caspar. Wärmetransport im Wettbewerb zu dislozierter Wärmeerzeugung. `http://www.eneff-stadt.info/fileadmin/media/Projektbilder/Neue_Technologien/Kostenguenstiger_Fernwaermetransport_Mvv/Abschlussbericht__Teilprojekt_1_Waermetransport_im_Wettbewerb_2014_10_07.pdf`, 2013. (accessed January 19, 2017).

[LB17] Michał Leśko and Wojciech Bujalski. Modeling of district heating networks for the purpose of operational optimization with thermal energy storage. *Archives of Thermodynamics*, 38(4):139–163, 2017.

[LBW04] Helge V. Larsen, Benny Bøhm, and Michael Wigbels. A comparison of aggregated models for simulation and operational optimisation of district heating networks. *Energy Conversion and Management*, 45(7-8):1119–1139, 2004.

[LM17] Emmanouil Loukarakis and Pierluigi Mancarella. A sequential programming method for multi-energy districts optimal power flow. In *2017 IEEE Manchester PowerTech*, pages 1–6, 2017.

[LØC+16] Henrik Lund, Poul Alberg Østergaard, David Connolly, Iva Ridjan, Brian Vad Mathiesen, Frede Hvelplund, Jakob Zinck Thellufsen, and Peter Sorknæs. Energy storage and smart energy systems. *International Journal of Sustainable Energy Planning and Management*, 11:3–14, 2016.

[LWLL17] Ping Li, Haixia Wang, Quan Lv, and Weidong Li. Combined heat and power dispatch considering heat storage of both buildings and pipelines in district heating system for wind power integration. *Energies*, 10(12):893, 2017.

[LWS+16] Zhigang Li, Wenchuan Wu, Mohammad Shahidehpour, Jianhui Wang, and Boming Zhang. Combined heat and power dispatch considering pipeline energy storage of district heating network. *IEEE Transactions on Sustainable Energy*, 7(1):12–22, 2016.

[LWW+14] Henrik Lund, Sven Werner, Robin Wiltshire, Svend Svendsen, Jan Eric Thorsen, Frede Hvelplund, and Brian Vad Mathiesen. 4th Generation District Heating (4GDH): Integrating smart thermal grids into future sustainable energy systems. *Energy*, 68:1–11, 2014.

[LWW+16] Zhigang Li, Wenchuan Wu, Jianhui Wang, Boming Zhang, and Taiyi Zheng. Transmission-constrained unit commitment considering combined electricity and district heating networks. *IEEE Transactions on Sustainable Energy*, 7(2):480–492, 2016.

[MAAFFH14] Moein Moeini-Aghtaie, Ali Abbaspour, Mahmud Fotuhi-Firuzabad, and Ehsan Hajipour. A decomposed solution to multiple-energy carriers optimal power flow. *IEEE Transactions on Power Systems*, 29(2):707–716, 2014.

[MC17] Hossein Mostafaei and Pedro M. Castro. Continuous-time scheduling formulation for straight pipelines. *AIChE Journal*, 63(6):1923–1936, 2017.

[McC76] Garth P. McCormick. Computability of global solutions to factorable nonconvex programs: Part I. Convex underestimating problems. *Math. Program.*, 10:147–175, 1976.

[NNM00] Henrik Aalborg Nielsen, Torben Skov Nielsen, and Henrik Madsen. *Load scheduling for decentralized CHP plants*. Informatics and Mathematical Modelling, Technical University of Denmark, DTU, Lyngby, 2000.

[Opp15] Thomas Oppelt. *Modell zur Auslegung und Betriebsoptimierung von Nah- und Fernkältenetzen*. Dissertation, TU Chemnitz, Chemnitz, 2015.

[Ran16] Tyler Ransom. An introduction to the Julia programming language, December 30, 2016.

[RG94] Ramesh Raman and Ignacio E. Grossmann. Modelling and computational techniques for logic based integer programming. *CPC VIII*, 18(7):563–578, 1994.

[ROGJ17] Dmytro Romanchenko, Mikael Odenberger, Lisa Göransson, and Filip Johnsson. Impact of electricity price fluctuations on the operation of district heating systems: A case study of district heating in Göteborg, Sweden. *Applied Energy*, 204:16–30, 2017.

[RP04] Ronald Rejowski and José M. Pinto. Efficient MILP formulations and valid cuts for multiproduct pipeline scheduling. *CPC VIII*, 28(8):1511–1528, 2004.

[Sah96] Nikolaos V. Sahinidis. BARON: a general purpose global optimization software package. *Journal of Global Optimization*, 8(2):201–205, 1996.

[SDEW12] Stephan Schweikardt, Michael Didycz, Felix Engelsing, and Katharina Wacker. Sektoruntersuchung Fernwärme: Abschlussbericht gemäß § 32e GWB, 2012.

[SFT+05] Guillaume Sandou, Stephane Font, Sihem Tebbani, Arnaud Hiret, and Christian Mondon. Predictive control of a complex district heating network. In *Proceedings of the 44th IEEE Conference on Decision and Control, and the European Control Conference*, 2005.

[SLM+17] Gerald Schweiger, Per-Ola Larsson, Fredrik Magnusson, Patrick Lauenburg, and Stéphane Velut. District heating and cooling systems – framework for Modelica-based simulation and dynamic optimization. *Energy*, 137:566 – 578, 2017.

[Sol06] Solar Energy Laboratory. TRNSYS 16 - Mathematical reference, 2006.

[SP08] Fabricio Salgado and Pedro Pedrero. Short-term operation planning on cogeneration systems: A survey. *Electric Power Systems Research*, 78(5):835–848, 2008.

[Sta15a] Stadtwerke Kiel. TAB Fernwärmenetze. `https://www.stadtwerke-kiel.de/swk/media/pdf/produkte_1/fernwaerme_2/stadtwerke-kiel-anschlussbedingungen-tab-fernwaerme.pdf`, 2015. (accessed July 20, 2019).

[Sta15b] Stadtwerke Kiel. Für die Energie von morgen: Unser Unternehmen im Überblick. `https://www.stadtwerke-kiel.de/swk/media/pdf/unternehmen_1/unternehmensbroschueren_1/stadtwerke-kiel-unternehmensbroschueren-unternehmen-2015.pdf`, 2015. (accessed February 21, 2019).

[Sta16] Stadtwerke Kiel. Küstenkraftwerk K.I.E.L. Kiels intelligente Energie-Lösung. `https://www.stadtwerke-kiel.de/swk/media/pdf/unternehmen_1/unternehmensbroschueren_1/stadtwerke-kiel-unternehmensbroschueren-kuestenkraftwerk-kiel.pdf`, 2016. (accessed June 17, 2017).

[TCM12] João P. Teles, Pedro M. Castro, and Henrique A. Matos. Global optimization of water networks design using multiparametric disaggregation. *Computers & Chemical Engineering*, 40:132–147, 2012.

[Trö99] Stefan Tröster. *Zur Betriebsoptimierung in Kraft-Wärme-Kopplungssystemen unter Berücksichtigung der Speicherfähigkeit des Fernwärmenetzes*. Dr.-Ing., Universität Dortmund, Dortmund, 1999.

[VD15] Mattias Vesterlund and Jan Dahl. A method for the simulation and optimization of district heating systems with meshed networks. *Energy Conversion and Management*, 89:555–567, 2015.

[vFR+17] Bram van der Heijde, Marcus Fuchs, Carles Ribas Tugores, Gerald Schweiger, Kevin Sartor, Daniele Basciotti, Dirk Müller, Christoph Nytsch-Geusen, Michael Wetter, and Lieve Helsen. Dynamic equation-based thermo-hydraulic pipe model for district heating and cooling systems. *Energy Conversion and Management*, 151:158–169, 2017.

[VTD17] Mattias Vesterlund, Andrea Toffolo, and Jan Dahl. Optimization of multi-source complex district heating network, a case study. *Energy*, 126:53–63, 2017.

[WB06] Andreas Wächter and Lorenz T. Biegler. On the implementation of an interior-point filter line-search algorithm for large-scale nonlinear programming. *Math. Program.*, 106(1):25–57, 2006.

[WBS05] Michael Wigbels, Benny Bøhm, and K. Sipilä. *Dynamic Heat Storage Optimisation and Demand Side Management*. IEA Annex VII. 2005:8DHC-05.06., 2005.

[Wer17] Sven Werner. International review of district heating and cooling. *Energy*, 137:617–631, 2017.

[WL18] Jonas Benjamin Weber and Ulf Lorenz. Algorithmic system design of thermofluid systems. `http://arxiv.org/pdf/1806.09939`, 2018. (accessed January 22, 2019).

[ZFZC18] Qing Zeng, Jiakun Fang, Baohua Zhang, and Zhe Chen. The coordinated operation of electricity, gas and district heating systems. *Energy Procedia*, 145:307–312, 2018.

[ZZZW18] Jinfu Zheng, Zhigang Zhou, Jianing Zhao, and Jinda Wang. Integrated heat and power dispatch truly utilizing thermal inertia of district heating network for wind power integration. *Applied Energy*, 211:865–874, 2018.

Own publications and conference contributions

[BBFM13] Shrikant Bhat, Naveen Bhutani, Carsten Franke, and Lennart Merkert. Keeping it cool: Optimal cooling systems design and management. *ABB Review*, 4:53–57, 2013.

[DMO+15] Marcel Dix, Lennart Merkert, Zied Ouertani, Markus Aleksy, and Reinhard Bauer. Simulation and re-engineering of industrial services: A case study from an industrial equipment manufacturer. *Proceedings - 2015 10th International Conference on P2P, Parallel, Grid, Cloud and Internet Computing, 3PGCIC 2015*, 2015.

[HCS+16] Martin Hollender, Moncef Chioua, Jan Schlake, Lennart Merkert, and Heiko Petersen. KPI-based process operation: Management of highly automated processes. *atp-edition*, 9:52–61, 2016.

[HM19] Iiro Harjunkoski and Lennart Merkert. Solving large-scale unit commitment problems. In *ETG-Kongress 2019 - Das Gesamtsystem im Fokus der Energiewende (ETG-Kongress 2019)*, Esslingen am Neckar, Germany, 2019.

[HMH+15] Iiro Harjunkoski, Lennart Merkert, Hubert Hadera, A. Shemeikka, D. Gajic, and Luca Onofri. Dicing the load: Flexible production saves energy costs. *ABB Review*, 3:18–24, 2015.

[HMS18] Iiro Harjunkoski, Lennart Merkert, and Jan Schlake. Demand side response (DSR) for improving resource efficiency beyond single plants. In Stefan Krämer and Sebastian Engell, editors, *Resource Efficiency of Processing Plants*, pages 293–316. Wiley-VCH, Weinheim, 2018.

[MC20] Lennart Merkert and Pedro M. Castro. Optimal scheduling of a district heat system with a combined heat and power plant considering pipeline dynamics. *Industrial & Engineering Chemistry Research*, 59(13):5969–5984, 2020.

[MF17] Lennart Merkert and Rüdiger Franke. Multi-energy site optimization. In *2016 European Control Conference, ECC 2016*, Aalborg, Denmark, 2017.

[MH17] Lennart Merkert and Iiro Harjunkoski. Integrating energy optimization and production scheduling in energy-intensive industries. In Georgios M. Kopanos, Pei Liu, and Michael C. Georgiadis, editors, *Advances in Energy Systems Engineering*, pages 601–620. Springer International Publishing, Cham, 2017.

[MH19] Lennart Merkert and Sören Hohmann. Globally optimal short-term unit commitment and dispatch for combined heat and power generation units connected to district heating grids. In Bernard Fortz and Martine Labbé, editors, *Operations Research Proceedings 2018*, pages 175–181, Cham, 2019. Springer International Publishing.

[MHH19] Lennart Merkert, Ashvar Abdoul Haime, and Sören Hohmann. Optimal scheduling of combined heat and power generation units using the thermal inertia of the connected district heating grid as energy storage. *Energies*, 12(2), 2019.

[MHI+15] Lennart Merkert, Iiro Harjunkoski, Alf Isaksson, Simo Säynevirta, Antti Saarela, and Guido Sand. Scheduling and energy - industrial challenges and opportunities. *Computers and Chemical Engineering*, 72:183–198, 2015.

[MKV+19] Lennart Merkert, Andreas Kohler, Annette Vogt, Kim Listmann, and Sören Hohmann. Increasing flexibility of CHP operations using heating grid dynamics. In *ETG-Kongress 2019 - Das Gesamtsystem im Fokus der Energiewende (ETG-Kongress 2019)*, Esslingen am Neckar, Germany, 2019.

[MLH19] Lennart Merkert, Kim Listmann, and Sören Hohmann. Optimization of thermo-hydraulic systems using multiparametric delay modeling. *Energy*, 189:116125, 2019.

[MMB17] Lennart Merkert, Miroslav Mendros, and Andreas Berthold. Optimal unit commitment in microgrids with frequency reserve constraints. *Proceedings of the 2017 IEEE 14th International Conference on Networking, Sensing and Control, ICNSC 2017*, 2017.

[MSIS16] Lennart Merkert, Antti Saarela, Alf Isaksson, and Simo Säynevirta. Scheduling thermo-mechanical pulp (TMP) production with respect to volatile energy cost. *2016 Control Systems Conference Stockholm*, 2016.

[SBM17] Sleman Saliba, Reinhard Bauer, and Lennart Merkert. Applications of virtual power plants - demand response at industrial sites. *PowerGen 2017*, 2017.

[SMB16] Sleman Saliba, Lennart Merkert, and Reinhard Bauer. Demand response in industrial sites - energy cost reduction using mathematical optimization. *OR 2016: International Conference on Operations Research Hamburg*, 2016.

Supervised master theses and internships

[Bau18a] Julian Bauer. Modeling of district heating grids. Internship, ABB AG, 2018.

[Bau18b] Julian Bauer. Modeling, simulation and non-linear optimization of a district heating grid. Master thesis, TH Mittelhessen University of applied Sciences, 2018.

[Bro17] Julia Brockschmidt. District heating modelling for virtual power plant optimisation. Master thesis, Faculty of Energy and Management, Technical University of Clausthal, 2017.

[Hai18] Ashvar Abdoul Haime. Optimal scheduling of combined heat and power generation units using the thermal inertia of the connected district heating grid as energy storage. Master thesis, Electrical and electronic section, Ecole Polytechnique Federale de Lausanee (EPFL), 2018.

[Lor19] Tobias Lorenz. Optimization of district heating grids. Internship, ABB AG, 2019.

Karlsruher Beiträge zur Regelungs- und Steuerungstechnik
Institut für Regelungs- und Steuerungssysteme
Karlsruher Institut für Technologie (ISSN 2511-6312)

Band 01 Diehm, Gunter
Identifikation des menschlichen Bewegungsverhaltens auf der Basis von Primitiven. 2017
ISBN 978-3-7315-0608-9

Band 02 Flad, Michael
Kooperative Regelungskonzepte auf Basis der Spieltheorie und deren Anwendung auf Fahrerassistenzsysteme. 2017
ISBN 978-3-7315-0610-2

Band 03 Eckert, Marius
Modellbasierte Identifikation fraktionaler Systeme und ihre Anwendung auf die Lithium-Ionen-Zelle. 2017
ISBN 978-3-7315-0690-4

Band 04 Krebs, Stefan
Intervallbeobachter für lineare parametervariante Systeme und deren Anwendung auf die Asynchronmaschine. 2019
ISBN 978-3-7315-0857-1

Band 05 Kaspar, Stephan
Fahrdynamikuntersuchungen eines Elektrofahrzeugs mit Einzelrad-Hinterradantrieb. 2019
ISBN 978-3-7315-0916-5

Band 06 Sauter, Patrick S.
Modellierung und zentrale prädiktive Regelung von multimodalen Energieverteilnetzen. 2019
ISBN 978-3-7315-0963-9

Band 07 Kupper, Martin
Verteilte Zustandsschätzung fraktionaler Systeme und ihre Anwendung auf Lithium-Ionen-Batteriesysteme. 2019
ISBN 978-3-7315-0971-4

Die Bände sind unter www.ksp.kit.edu als PDF frei verfügbar oder als Druckausgabe bestellbar.

Band 08 Merkert, Lennart
Optimal Scheduling of Combined Heat and Power Generation Considering Heating Grid Dynamics. 2021
ISBN 978-3-7315-1056-7
